Colorimetric Determination of Nitrate Plus Nitrite in Water by Enzymatic Reduction, Automated Discrete Analyzer Methods

By Charles J. Patton and Jennifer R. Kryskalla

Chapter 8
Section B, Methods of the National Water Quality Laboratory
Book 5, Laboratory Analysis

Techniques and Methods 5–B8

U.S. Department of the Interior
U.S. Geological Survey

U.S. Department of the Interior
KEN SALAZAR, Secretary

U.S. Geological Survey
Marcia K. McNutt, Director

U.S. Geological Survey, Reston, Virginia: 2011

For more information on the USGS—the Federal source for science about the Earth, its natural and living resources, natural hazards, and the environment, visit http://www.usgs.gov or call 1–888–ASK–USGS.

For an overview of USGS information products, including maps, imagery, and publications, visit http://www.usgs.gov/pubprod

To order this and other USGS information products, visit http://store.usgs.gov

Contents

Figures

Tables

Conversion Factors

[SI to Inch/Pound]

Multiply	By	To obtain
Length		
centimeter (cm)	3.94×10^{-1}	inch
micrometer (μm)	3.94×10^{-5}	inch
millimeter (mm)	3.94×10^{-2}	inch
nanometer (nm)	3.94×10^{-8}	inch
Mass		
gram (g)	3.53×10^{-2}	ounce, avoirdupois
milligram (mg)	3.53×10^{-5}	ounce, avoirdupois
Volume		
liter (L)	2.64×10^{-1}	gallon
liter (L)	3.38×10^{1}	ounce, fluid
microliter (μL)	2.64×10^{-7}	gallon
milliliter (mL)	2.64×10^{-4}	gallon
Concentration		
milligrams nitrate nitrogen per liter (mg NO_3^--N/L)	7.14×10^{1}	micromoles nitrate per liter (NO_3^-, μM)
milligrams nitrite nitrogen per liter (mg NO_2^--N/L)	7.14×10^{1}	micromoles nitrite per liter (NO_2^-, μM)
molar (M, moles/L)	1.00×10^{3}	millimolar (mM)
molar (M, moles/L)	1.00×10^{6}	micromolar (μM)

Temperature in degrees Celsius (°C) may be converted to degrees Fahrenheit (°F) as follows:

$$°F = 9/5 \,(°C) + 32$$

Acronyms and Abbreviations

≈	Approximately equal to
≥	Greater than or equal to
ASTM	American Society for Testing and Materials
AtNaR2™	Recombinant nitrate reductase from *Arabidopsis thaliana* (Enzyme Commission # EC 1.7.1.1) expressed in *Pichia pastoris*
BQS	USGS Branch of Quality Systems
CdR	Cadmium reduction
CFA	Air-segmented continuous-flow analysis (or analyzer)
DA	Automated discrete analysis (or analyzer)
DI (water)	Deionized water piped throughout the NWQL. For purposes of nutrient analysis, NWQL DI water is equivalent to ASTM type I DI.
DOC	Dissolved organic carbon
DOM	Dissolved organic matter

DTPA	Diethylenetriaminepentaacetic acid
EC	Enzyme Commission; Swiss organization that assigns unique numerical identifiers to enzymes
EDTA	Ethylenediaminetetraacetic acid and its di-sodium salt
EPA	U.S. Environmental Protection Agency
FCA	Bottle type for nutrient samples that are filtered (0.45 µm), amended with sulfuric acid, and chilled at collection sites. FCA samples are shipped and stored at a nominal temperature of 4°C.
FCC	Bottle type for nutrient samples that are filtered (0.45 µm) and chilled at collection sites. FCC samples are shipped and stored at a nominal temperature of 4°C.
FW	Formula weight
HA	High-phenolic-content humic acid
IRL	NWQL interim reporting level (or limit)
LC	NWQL laboratory code
LCL	Lower control limit
LL	Low-level analytical range
LRL	NWQL laboratory reporting limit
LT-MDL	NWQL long-term method detection level (or limit)
M	Molar (molarity); unit of concentration equal to moles solute per liter of solution
MC	USGS NWIS sample medium code used to identify sample matrix types
MDL	Method detection level (or limit)
MOPS	3-N-morpholino-propansulfonic acid; $pK_a = 7.20$
MPV	Most probable value
MRL	Method reporting level (or limit)
NAD(P)H:NaR	Bispecific forms of nitrate reductase that can use either NADH or NADPH as their cofactor
NADH	Nicotinamide adenine dinucleotide in reduced form, a natural cofactor of YNaR1 and AtNaR2
NADH:NaR	Forms of nitrate reductase that can use only NADH as their cofactor
NADPH	Nicotinamide adenine dinucleotide phosphate in reduced form, the other natural cofactor of bispecific YNaR1 and of nitrate reductase from *Aspergillus* sp.
NADPH:NaR	Forms of nitrate reductase that can use only NADPH as their cofactor
NaR	Nitrate reductase in the generic sense
NaR1™	Nitrate reductase purified from corn (*Zea mays*) by NECi
NECi	The Nitrate Elimination Company, Lake Linden, MI 49945
NED	*N*-(1-Naphthyl)ethylenediamine; coupling reagent in Griess-reaction nitrite

	assays
NWIS	USGS National Water Information System
NWQL	USGS National Water Quality Laboratory
pK_a	$-\log_{10} K_a$, where K_a is the dissociation constant of a weak acid
QC	Quality control; used as an adjective in this report for materials analyzed to assure analytical results are within specified limits
RSD	Relative standard deviation
SAN	Sulfanilamide; diazotizing reagent in Griess-reaction nitrite assays
SL	Standard-level analytical range
SOP	Standard operating procedure
SR HA	Suwannee River humic acid isolate; a high-phenolic-content HA
TPC	Third-party check sample. A QC sample prepared with a certified stock analyte solution obtained from a source different from the one used to prepare calibrants.
UCL	Upper control limit
USGS	U.S. Geological Survey
v/v	Volume-to-volume
WCA	Bottle type for whole-water nutrient samples that are amended with sulfuric acid and chilled at collection sites. WCA samples are shipped and stored at a nominal temperature of 4°C.
WG	USGS NWIS medium code for groundwater (formerly medium code 6)
WS	USGS NWIS medium code for surface water (formerly medium code 9)
YNaR1™	Recombinant nitrate reductase from *Pichia angusta* (Enzyme Commission # EC 1.7.1.2) expressed in *Pichia pastoris* and purified to near homogeneity

[AtNaR2, NaR1, and YNaR1 are trade names of the Nitrate Elimination Company, Inc., Lake Linden, MI 49945]

Colorimetric Determination of Nitrate Plus Nitrite in Water by Enzymatic Reduction, Automated Discrete Analyzer Methods

By Charles J. Patton [1] and Jennifer R. Kryskalla [2]

Abstract

This report documents work at the U.S. Geological Survey (USGS) National Water Quality Laboratory (NWQL) to validate enzymatic reduction, colorimetric determinative methods for nitrate + nitrite in filtered water by automated discrete analysis. In these standard- and low-level methods (USGS I-2547-11 and I-2548-11), nitrate is reduced to nitrite with nontoxic, soluble nitrate reductase rather than toxic, granular, copperized cadmium used in the longstanding USGS automated continuous-flow analyzer methods I-2545-90 (NWQL laboratory code 1975) and I-2546-91 (NWQL laboratory code 1979). Colorimetric reagents used to determine resulting nitrite in aforementioned enzymatic- and cadmium-reduction methods are identical. The enzyme used in these discrete analyzer methods, designated AtNaR2 by its manufacturer, is produced by recombinant expression of the nitrate reductase gene from wall cress (*Arabidopsis thaliana*) in the yeast *Pichia pastoris*. Unlike other commercially available nitrate reductases we evaluated, AtNaR2 maintains high activity at 37°C and is not inhibited by high-phenolic-content humic acids at reaction temperatures in the range of 20°C to 37°C. These previously unrecognized AtNaR2 characteristics are essential for successful performance of discrete analyzer nitrate + nitrite assays (henceforth, DA-AtNaR2) described here.

Method detection levels (or limits; MDL) estimated for standard- and low-level DA-AtNaR2 nitrate + nitrite methods were 0.02 milligrams nitrogen per liter (mg-N/L) and 0.002 mg-N/L, respectively, which are comparable to 2010 NWQL long-term MDLs of the continuous-flow analyzer, cadmium-reduction methods (henceforth, CFA-CdR) they replace. Typically, reagent-water blanks for standard- and low-level DA-AtNaR2 nitrate + nitrite methods are one half MDL or less. Nitrate + nitrite concentration differences for between-day replicates were 3 percent or less at or above 5 times the MDL and were as great as 35 percent near the MDL. Typically, nitrate spike recoveries from reagent water, surface water, groundwater, and high-phenolic-content, humic-acid-amended reagent water were 100±20 percent.

In addition to operational details and performance benchmarks for these new DA-AtNaR2 nitrate + nitrite assays, this report also provides results of interference studies for common inorganic and organic matrix constituents at 1, 10, and 100 times their median concentrations in surface-water and groundwater samples submitted annually to the NWQL for nitrate + nitrite analyses. Paired t-test and Wilcoxon signed-rank statistical analyses of results determined by CFA-CdR methods and DA-AtNaR2 methods indicate that nitrate concentration differences between population means or sign ranks were either statistically equivalent to zero at the 95 percent confidence level ($p \geq 0.05$) or analytically equivalent to zero—that is, when $p < 0.05$, concentration differences between population means or medians were less than MDLs.

Introduction

Nitrate (NO_3^-) is one of the most universally determined anions in natural water and drinking water because it can promote eutrophication and is toxic to fetuses and young of livestock and humans at concentrations that exceed about 10 milligrams nitrogen per liter (mg-N/L) (U.S. Environmental Protection Agency, 1995). A thorough review of detection and determination methods for nitrate and nitrite (NO_2^-) in a variety of matrices is available elsewhere (Moorcroft and others, 2001). Some important references not cited in Moorcroft's review include one describing reduction of nitrate to nitrite with trivalent vanadium (Miranda, 2001), another on optimizing cadmium-reduction assays (Gal and others, 2004), a third documenting ferrous iron interference in the Griess colorimetric indicator reaction (Colman and Schimel, 2010a, b), and several pertaining to nitrate-reductase-based nitrate assays (Senn and Carr, 1976; Guevara and others, 1998; Mori 2000, 2001; Patton and others, 2002; MacKown and Weik, 2004; Pinto and others, 2005; Campbell and others, 2006).

[1] U.S. Geological Survey, National Water Quality Laboratory, Denver, Colo.

[2] Veterans Health Administration, VISN 21 Pharmacy Benefits Management Group, Reno, Nev.

Cadmium in various forms—such as electrolytically precipitated, "mossy" or "spongy," filings, granules, and filings or granules washed with solutions of mercury (II), silver (I), or copper (II) ions (Nydahl, 1976; Davison and Woof, 1978)—has long been the reducing agent of choice for colorimetric nitrate determinations. For example, copper-washed (copperized) cadmium granules packed into small columns (Wood and others, 1967) are prescribed in the longstanding U.S. Geological Survey (USGS) and U.S. Environmental Protection Agency (EPA) continuous-flow analyzer, cadmium-reduction (CFA-CdR) methods I-2545-90 and 353.2, respectively. Wire-in-tube cadmium reactors (Stainton, 1974; Willis, 1980; Willis and Gentry, 1987; Patton and Rogerson, 2007) and open-tubular cadmium reactors (Patton, 1983; Elliot and others, 1989; Zhang and others, 2000) are well known and effective alternatives to packed-bed reactors. A definitive study on continuous-flow cadmium reactors (Nydahl, 1976) demonstrated that reaction-stream pH in the range of 7.0 to 8.5 is required for near-quantitative reduction of nitrate to nitrite with only minor (less than 3 percent) reduction of nitrite to lower oxidation species. Long-term reactor stability also depends critically on including reagents in the analytical stream that form strong complexes with cadmium (II) ions—imidazole or ammonium chloride, typically. Without such reagents, cadmium (II) ions formed during reactions between cadmium and nitrate, dissolved oxygen, or both would precipitate as hydroxides on cadmium surfaces and deactivate them.

Despite their long predominance as reducing agents of choice for colorimetric nitrate determinations in water, flow-through cadmium reactors are difficult to prepare and activate, pose health risks to analysts and waste stream processors, increase waste stream disposal costs, and are incompatible with discrete analyzers. These drawbacks motivated the National Water Quality Laboratory (NWQL) to explore commercially available nitrate reductase enzymes as soluble, nontoxic replacements for cadmium. Success of preliminary work (Patton and others, 2002) provided motivation and continued institutional support for further studies in which we investigated and validated two other nitrate reductase enzymes as direct replacements for cadmium in USGS-approved colorimetric nitrate + nitrite assays. The automated discrete analyzer standard- and low-level enzymatic-reduction, colorimetric nitrate + nitrite assays described in the sections that follow are the end products of this multiyear research effort.

Purpose and Scope

This report describes new, enzymatic-reduction methods for colorimetric nitrate + nitrite determinations in surface water and groundwater on automated discrete analyzer (DA) instrument platforms. In this report, we provide the following information to NWQL customers and other USGS data users who interpret or report nitrate concentration data and to analysts at the NWQL and elsewhere who need to implement these methods and routinely operate them:

1. Graphical and statistical analysis of paired analytical data demonstrating equivalence of nitrate + nitrite concentrations determined by these new methods (I-2547-11 and I-2548-11) and by time-honored USGS CFA-CdR methods I-2545-90 and I-2546-91;

2. Operational details and performance benchmarks for these new discrete-analyzer, AtNaR2-reduction (DA-AtNaR2) methods, including method detection levels (or limits, MDLs), blank levels, between-day precision, and spike recovery from reagent water, surface water, groundwater, and high-phenolic-content, humic-acid-amended, reagent water, and;

3. Summaries of experiments demonstrating negligible interference in enzymatic and colorimetric assay reaction steps by common surface-water and groundwater matrix constituents such as major and minor ions and humic substances over a reaction temperature range of 5°C to 37°C.

This report focuses on development and validation of standard- and low-level discrete analyzer nitrate + nitrite assays using AtNaR2 nitrate reductase (Skipper and others, 2001; Campbell and others, 2006) and its cofactor, β-nicotinamide adenine dinucleotide, reduced form (NADH). The NWQL analytical services sample stream was the source of seasonally and geographically diverse surface-water and groundwater samples that we used to demonstrate capability and validate these new methods. These samples had nitrate concentrations ranging from hundredths to tens of milligrams nitrogen per liter. Specifically, we used these new enzymatic reduction methods to analyze nitrate + nitrite in subsets of samples originally submitted to the NWQL for analysis by USGS-approved cadmium-reduction methods, and we then compared results of the new analyses with the previous results. This approach is practical, cost effective, and would clearly indicate bias, if any, in nitrate + nitrite concentrations determined in real samples during routine operation by USGS-approved methods and new methods. Data and statistical analysis supporting established 30-day holding times for nitrate + nitrite in filtered and filtered-acidified water samples are published elsewhere (Patton and Truitt, 1995; Patton and Gilroy, 1998).

Analytical Methods

1. Application

The subject new methods listed in *table 1* are suitable for determination of nitrate + nitrite in filtered (FCC bottle type) and filtered-acidified (FCA bottle type) water samples. They also are applicable to whole-water-acidified (WCA bottle type) Samples that are laboratory filtered prior to analysis. They are direct replacements for longstanding USGS and EPA colorimetric nitrate + nitrite methods and differ from them only in the reagents used to reduce nitrate to nitrite (nontoxic, soluble nitrate reductase replaces toxic, granular, copperized cadmium) prior to colorimetric nitrite determination with Griess reagents. Like cadmium-reduction methods, these enzymatic-reduction methods are intended for surface-water and groundwater matrices (NWIS medium codes WG and WS, formerly 6 and 9). Seawater, brines, leachates, potassium chloride soil extracts, landfill effluents, and other nonconforming matrices should not be submitted for analyses without prior consultation with the NWQL. Such matrices do not match those of calibrants and quality-control samples and therefore might produce incorrect analytical results. Nominal analytical ranges for standard- and low-concentration methods are 0.04 to 5.00 mg-N/L and 0.008 to 1.00 mg-N/L, respectively.

2. Method Summaries and Analytical Considerations

NADH:nitrate reductase, Enzyme Commission number EC 1.7.1.1, hereafter designated AtNaR2 (Skipper and others, 2001; Campbell and others, 2006), requires NADH as its electron donating reagent (cofactor). AtNaR2 is produced through recombinant expression of the nitrate reductase gene from a land plant commonly known as "wall cress" (*Arabidopsis thaliana*) in the yeast *Pichia pastoris*. AtNaR2 is a proprietary product of the Nitrate Elimination Company (NECi), Lake Linden, Mich. In phosphate or 3-N-morpholino-propansulfonic acid (MOPS) buffers in the pH range of 7 to 8, AtNaR2 quantitatively reduces nitrate (NO_3^-) to nitrite (NO_2^-), as shown in equation 1.

$$NO_3^- + NADH + H^+ \xrightarrow[(pH=7-8)]{AtNaR2} NO_2^- + NAD^+ + H_2O \quad (1)$$

In accordance with the colorimetric reaction scheme below, resultant nitrite plus any nitrite present in the sample prior to enzymatic reduction diazotizes with sulfanilamide at pH ≈ 1. The p-diazonium sulfanilamide thus formed subsequently reacts with N-(1-Naphthyl)ethylenediamine (Bratton-Marshall variant of the Griess reaction) to form a pink, azo dye with an absorption maximum at 543 nm (Bratton and Marshall, 1939; Bendschneider and Robinson, 1952; Fox, 1979, 1985; Pai and others, 1990).

3. Interferences and Temperature Effects

Large buffer-to-sample ratios used in methods I-2547-11 and I-2548-11 mitigate the potential reduction-step and colorimetric-step interferences listed below.

3.1 Any particles in assays (turbidity) introduced by samples, reagents, or both scatter light during photometric measurements. Such turbidity contributes to chromophore absorbance and can cause high bias in analytical results. Discernible turbidity in samples or colorimetric reagents, therefore, should be removed by filtration (0.45-μm or 0.2-μm polyethersulfone or nylon) prior to analytical determinations.

Table 1. Laboratory, parameter, and method codes for U.S. Geological Survey automated discrete analyzer, enzymatic reduction, standard-level (I-2547-11) and low-level (I-2548-11) nitrate + nitrite determination methods.

[FCC, filtered chilled container; μm, micrometer; FCA, filtered, chilled, acidified container; mL, milliliter]

Description	Codes			Bottle type
	Laboratory	Parameter	Method	
Nitrate + nitrite, as N, colorimetry, DA, enzymatic reduction-diazotization, filtered (method I-2547-11)	3156	00631	RED01	FCC[1]
Nitrate + nitrite, as N, colorimetry, DA, enzymatic reduction-diazotization, filtred, low-level (method I-2548-11)	3157	00631	RED02	FCC[1]
Nitrate + nitrite, as N, colorimetry, DA, enzymatic reduction-diazotization, filtered acidified (method I-2547-11)	3222	00631	RED03	FCA[2]

[1]FCC samples must be processed through 0.45-μm filters and chilled at collection sites.
[2]FCA samples must be processed through 0.45-μm filters, chilled, and amended with 1 mL of 4.5 N H_2SO_4 solution (U.S. Geological Survey water-quality field supply number Q438FLD) per 120 mL of sample at collection sites.

3.2 High concentrations of certain transition- and heavy-metal ions can inhibit nitrate reductase to varying extents. Ethylenediaminetetraacetic acid (EDTA) forms strong complexes with many metal ions and effectively minimizes interference by these potential sample matrix constituents that might otherwise hinder quantitative reduction of nitrate to nitrite. Eight of the more abundant metal ions in water analyzed at the NWQL at up to 100 times their median annual concentrations affected nitrate recovery by less than ±2 percent. See the section titled "Analytical Performance and Comparative Results" for additional details.

3.3 Sulfate, chloride, and bromide at up to 100 times their median concentration in samples analyzed at the NWQL annually affected nitrate recovery by less than ±2 percent. Perchlorate at concentrations up to 5 mg/L affected nitrate recovery by less than 2 percent. See the section "Analytical Performance and Comparative Results" for additional details.

3.4 High-phenolic-content humic substances (HAs) are matrix constituents in perhaps 15 percent of water samples received for nitrate analysis at the NWQL annually. Concentrations of HA up to 20 mg/L do not inhibit AtNaR2 in the reduction reaction temperature range of 10°C to 37°C. However, for other commercially available nitrate reductase enzymes that we evaluated, HA inhibition was negligible only in the temperature range of 10°C to 20°C. Above 20°C, HA inhibition increased continuously in direct proportion to reduction reaction temperature and HA concentration (see section "Effects of Temperature and Dissolved Organic Matter on AtNaR2 Activity").

3.5 NADH inhibits the Griess indicator reaction (Patton and others, 2002, table 1; Moody and Shaw, 2006). Quantitative reduction of nitrate to nitrite with minimum Griess reaction inhibition occurs when initial NADH concentration is in two-fold molar excess to that of a method's maximum nitrate concentration in the reaction medium. Methods I-2547-11 and I-2548-11 conform to this initial NADH concentration condition.

3.6 AtNaR2 and other nitrate reductases we evaluated promote oxidation of NADH to NAD^+ even in the absence of nitrate. Separate AtNaR2 and NADH reagents used in assays described here eliminate the possibility of this potentially reagent-limiting side reaction. If a mixed AtNaR2-NADH reagent were required—because of analytical platform limitations, perhaps—its useful lifetime would be less than 2 hours.

3.7 Norwitz and Keliher (1985, 1986) systematically assessed inorganic and organic interferences for the Griess indicator reaction. Colman and Schimel (2010a, b) recently reported that Fe (II) at or above 10 mg/L suppresses the Griess reaction. According to these authors, replacing ethylenediaminetetraacetic acid (EDTA) with diethylenetriaminepentaacetic acid (DTPA) in nitrate assay buffers eliminates this interference. Although Fe (II) concentrations of 10 mg/L or more are unlikely to occur in surface water and groundwater, analysts applying these methods to high-iron soil extracts, acid mine drainage water, or pore water from low-oxygen bed sediments should be aware of this potential interference and its remedy.

3.8 A number of metal cations are minor nitrite indicator reaction inhibitors (see *fig. 1*). Group II (alkaline earth) cations produce the largest effects. Calcium ions reduce the yield of indicator reaction chromophore the most—about 5 percent at NWQL-median concentrations—but barium ions are the most potent indicator reaction suppressor on a molar basis.

3.9 The inverse relationship between reaction temperature in the range of 10°C to 50°C and formation rate and yield of the Griess indicator reaction chromophore evident in *figure 2* results from thermal instability of nitrous acid and diazonium intermediates in the Griess reaction (Noller, 1966).

4. Instrumentation

We developed automated DA soluble AtNaR2-reduction nitrate + nitrite methods using a Kone Aquakem 600™ analyzer (Thermo Fisher Diagnostics, Fremont, Calif.). Basic operation of the Aquakem 600™ DA can be understood with reference to *figure 3* and the text that follows.

On startup, cuvette segments—linear arrays of 12 cells in which individual tests or dilutions take place and through which absorbance is measured—move from the cuvette loader into available incubator slots. The incubator hub rotates to align cuvette segments with sample- or reagent-dispensing alleys as appropriate during the analytical cycle. Sample and reagent "disks" that hold sample segments and reagent containers are thermostatted at 10°C and 4°C, respectively. Precisely controlled rotation of these disks aligns the appropriate sample or reagent with dispensing arms during operation. As cuvette segments move sequentially through sample- and reagent-dispensing alleys, individual cells are aligned with high-precision dispensers attached to robotic arms. Stirrers on another set of robotic arms mix cell contents after each dispense cycle. The robotic arms return dispenser needles and stirrer blades to wash stations for thorough rinsing after each dispense/mix operation. Between dispensing operations, cuvette segments return to the incubator where programmed reaction times up to 60 minutes occur. The incubator, dispensing alleys, and photometer module are thermostatted at 37°C. Liquids dispensed into cells equilibrate to 37°C during the course of analyses. At the end of the three programmed incubations (see *table 2*), cuvette segments exit the incubator one at a time and enter the photometer module where sequential measurements of absorbance in each cell occurs. The DA ejects cuvette segments into a waste compartment positioned below the photometer after absorbance measurements are complete.

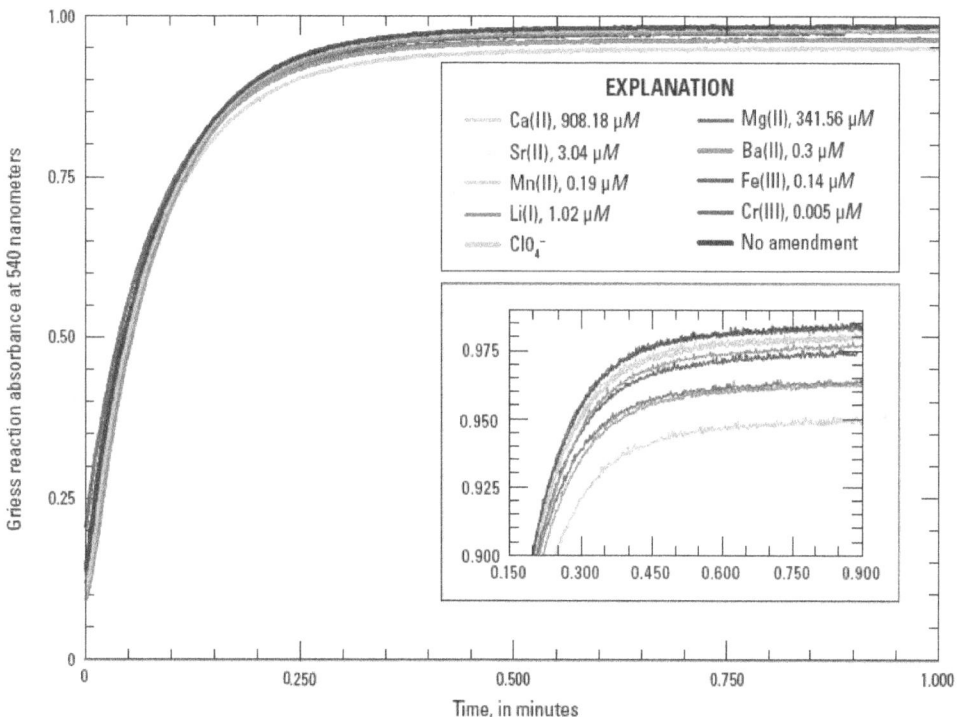

Figure 1. Kinetics effects of diverse metal ions and perchlorate on the Griess reaction colorimetric nitrite assay. (μM, micromolar)

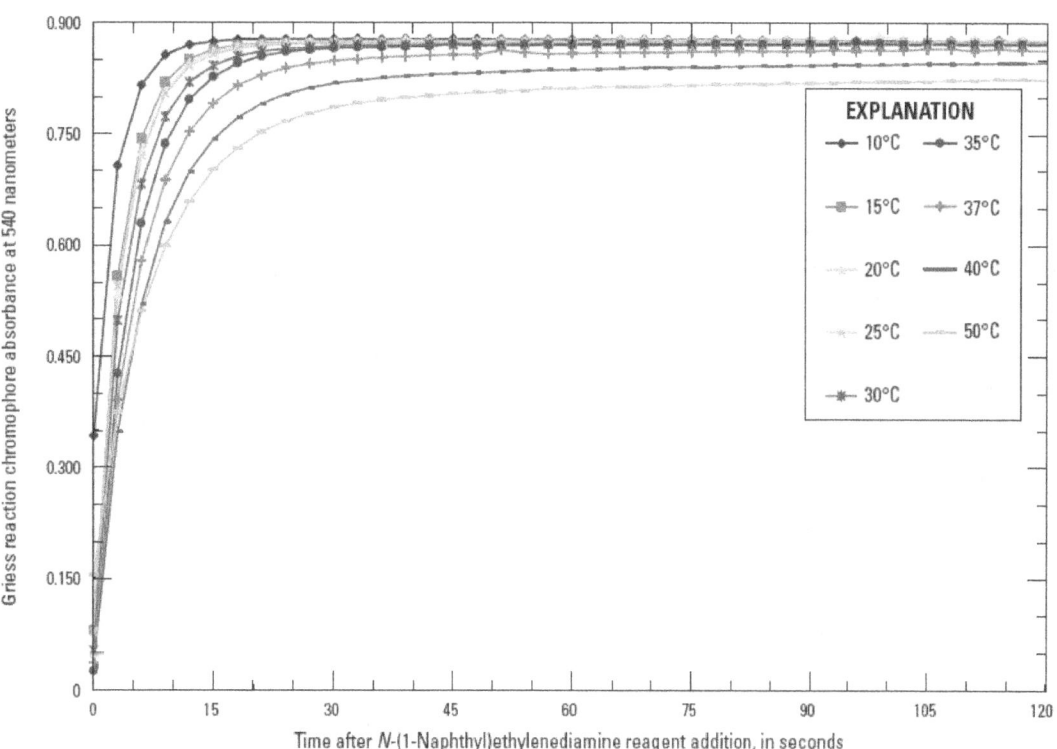

Figure 2. Effect of temperature on kinetics and yield of the Griess reaction nitrite assay. (°C, degrees Celsius)

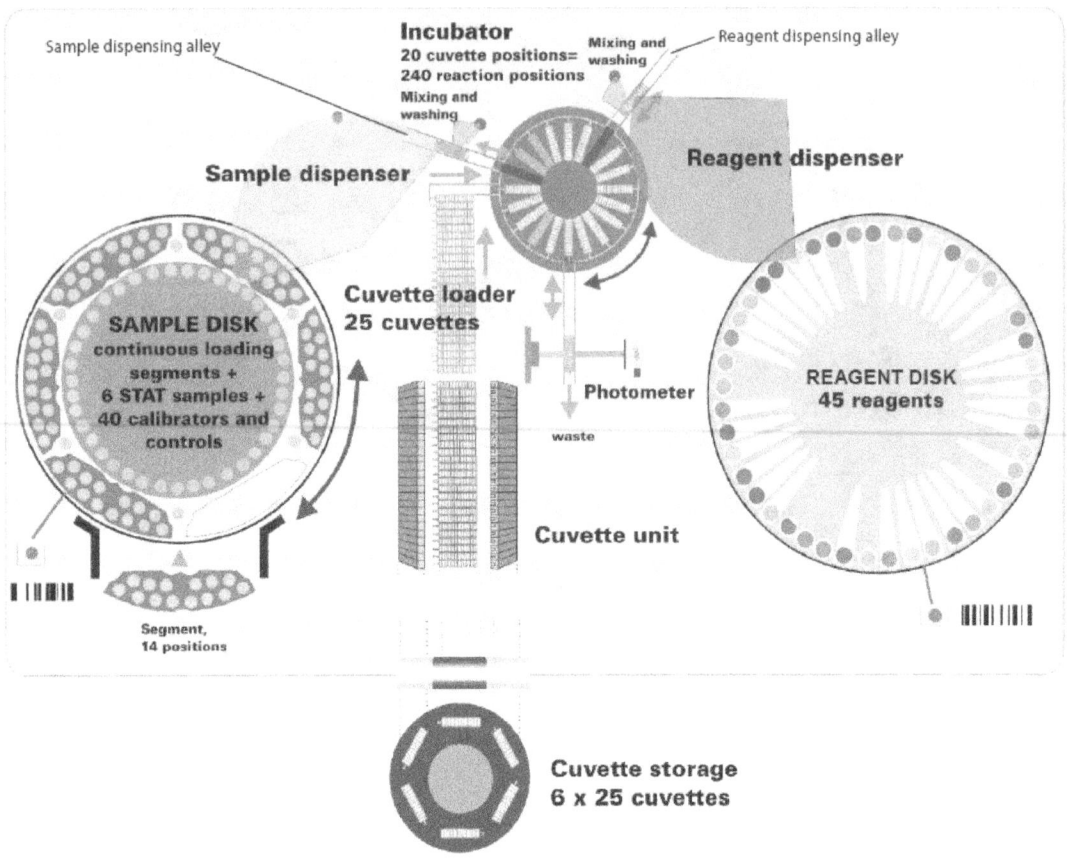

Figure 3.   Functional diagram of the Aquakem 600™ automated discrete analyzer. (Image courtesy of Thermo Scientific.)

The Aquakem 600™ measures absorbance with a "dual-beam-in-time" filter photometer (Ingle and Crouch, 1988). This photometer design compensates for wavelength-dependent light-source intensity and detector sensitivity as well as light source flicker and drift. Aquakem 600™ software provides two methods to record and correct minor contributions to assay absorbance caused by turbidity and cuvette imperfections. The first, termed side-wavelength correction, involves photometric measurement of finished assays at the wavelength where chromophore absorbance is maximum ($\lambda_{max} \approx 540$ nm) and at a second wavelength where chromophore absorbance is negligible (700 nm). The difference between absorbance measured at 540 nm and 700 nm yields chromophore absorbance corrected for light scattering effects. The correction works because light scattering in the wavelength range of 540–700 nm is nearly constant. The second method, termed reagent-blank correction, involves (1) measuring the intermediate assay absorbance after adding and mixing the first reagent sulfanilamide (SAN) at λ_{max}, (2) adding the second, color forming reagent N-(1-Naphthyl)ethylenediamine (NED), and (3) measuring finished assay absorbance again at λ_{max} after chromophore formation is complete. Here, correction and analytical absorbance measurement wavelengths are the same, but the small volume difference between measurements might

slightly overcorrect scattering effects. In practice, analytical results obtained with either correction method are the same within assay precision limits. We used side-wavelength correction during method development and validation. We have since implemented reagent-blank correction because the Kone software generates automatic warnings, report flags, and conditional branching from operator specified reagent-blank upper and lower absorbance limits. This functionality is not provided for side-wavelength absorbance corrections.

See *table 2* for DA operational protocols (test flows) used for standard- and low-level concentration AtNaR2-reduction nitrate + nitrite determinations. Because standard-level (SL) and low-level (LL) methods are identical except for a fivefold sample-volume increase in the latter, we combined their test flows in *table 2*. In this table, "extra" refers to aspirated volumes that are not delivered into cuvette cells. According to the DA manufacturer, dispensing with "extra" minimizes sample and reagent dilution during dispensing operations and thus improves precision of analytical results. "Extra" sample and reagent volumes are purged from dispensing needles during rinse cycles and collected in the analyzer waste stream. Additional details of Aquakem 600™ hardware and software are in the manufacturer's operation manual and NWQL Technical Operations Manual for Kone Aquakem 600™ DA (Schwab and others, 2009).

Table 2. Aquakem 600™ automated discrete analyzer operational protocols, termed test flows by the vendor, for standard-level (SL) and low-level (LL) AtNaR2-reduction nitrate + nitrite determination methods.

[μL, microliter; s, second; nm, nanometers; extra, aspirated volumes that are not delivered into reaction cells; NADH, nicotinamide adehine dinucleotide in reduced form; SAN, sulfanilamide; NED, *N*-(1-Naphthyl)ethylenediamine]

Dispensed liquid	Volume/extra (μL)	Incubation time (s)	Wavelength (nm)	
			Analytical	Side*
AtNaR2	55/10			
Sample, SL	5/25			
Sample, LL	25/25			
NADH	12/15			
		600		
SAN	25/10			
		120		
Reagent blank			540	
NED	25/10			
		120		
			540	700

*Optional.

5. Apparatus

We used EDP-*plus*™ electronic, digital pipets (Rainin Instruments, Oakland, Calif.) fitted with 10–100-μL, 100–1,000-μL, and 1,000–10,000-μL liquid ends as appropriate for most precision dispensing.

We configured the purpose-built, thermostatted, continuous flow reaction monitor used for interference studies and enzyme reaction rate experiments from components in our laboratory, including an OB-1 large-platform autosampler (Oregon Manufacturing Support, Malin, Oreg.), RFA-300 continuous flow analyzer modules (no longer in production), and a model TLC 40 temperature-controlled cuvette holder equipped with a magnetic stirring accessory (Quantam Northwest, Spokane, Wash.).

6. Reagent Preparation

This section provides detailed instructions for preparing enzymatic and colorimetric reagents used for standard- and low-level discrete analyzer assays. All references to deionized (DI) water refer to DI water piped throughout the NWQL. For purposes of nutrient analysis, NWQL DI water is comparable to ASTM type I DI water (American Society for Testing and Materials, 2001, p. 107–109). We triple rinsed all volumetric glassware and containers for reagent and calibrant storage with dilute ($\approx$ 5 percent v/v) hydrochloric acid and DI water just prior to use. We also triple rinsed reagent and calibrant storage containers with small portions of the solutions they were to contain before we filled them.

6.1 *Enzymatic reagents*

6.1.1 *Di-sodium ethylenediaminetetraacetic acid (EDTA), 25 millimolar (mM).*—Dissolve 9.3 g EDTA (FW = 372.24, Ultrapure grade) in approximately 800 mL DI water contained in a 1-L volumetric flask. Dilute the resulting solution to the mark with DI water, mix it well, and transfer it to a bottle where it is stable at room temperature for one year.

6.1.2 *Phosphate buffer (pH = 7.5).*—Dissolve 3.75 g potassium di-hydrogen phosphate (KH_2PO_4, FW = 136.1) and 1.4 g potassium hydroxide (KOH, FW = 56.11) in about 800 mL of DI water contained in a 1-L volumetric flask. Add 1 mL 25 m*M* EDTA and dilute the resulting solution to the mark with DI water; mix it well. Transfer this solution to a bottle where it is stable at room temperature for one year.

6.1.3 *Nitrate reductase from Arabidopsis thaliana, AtNaR2, EC #1.7.1.1.*—Remove the cap from a vial containing 3 units of freeze-dried AtNaR2 and add to it about 1 mL of the proprietary reconstitution buffer supplied with the enzyme. Alternatively, substitute 1 mL of pH 7.5 phosphate buffer (6.1.2). Recap the vial and invert it several times over the course of 30 minutes to speed dissolution of the freeze-dried enzyme.

NOTE: According to the enzyme manufacturer, 3 units of AtNaR2 dissolved in $\approx$1 mL of their proprietary reconstitution buffer are stable at or below -15°C for several months. NECi includes a squeezable plastic ampoule containing about 1 mL of this buffer with each 3-unit vial of AtNaR2.

6.1.4 *Working AtNaR2 reagent.*—Quantitatively transfer and dilute the dissolved enzyme concentrate in a 20-mL Kone reagent tube as follows:

- Carefully pour the dissolved enzyme concentrate from the vial in which it was reconstituted into the reagent tube.

- Use a digital pipet to dispense 1,000 μL of pH 7.5 phosphate buffer (6.1.2) into the empty enzyme vial.

- Recap the vial and invert it several times.

- Before removing the cap, tap it sharply with your finger to dislodge adherent droplets.

- Remove the cap and pour the resulting rinse solution into the reagent tube.

- Repeat steps 2–5 two more times, after which the reagent tube should contain 4 mL of enzyme concentrate in phosphate buffer.

- Add 16.0 mL of phosphate buffer (dispensing 8 mL twice from a digital pipet equipped with a 10-mL liquid end works well) into the reagent tube and recap it. Then mix the working reagent gently by repeated inversion. Working AtNaR2 enzyme reagent is stable at 2°C to 8°C for about 18 hours.

If a 20-mL batch of this reagent, which is sufficient for about 330 assays, cannot be used within a day, prepare a smaller volume—for example, 250 µL AtNaR2 concentrate diluted to 5 mL with pH 7.5 buffer—and store remaining 750 µL of AtNaR2 concentrate at or below -15°C for future use. Alternatively, remove remaining working AtNaR2 reagent from the analyzer and freeze it at or below -15°C.

6.1.5 β-*Nicotinamide adenine dinucleotide, reduced form, disodium salt (NADH) stock solution.*—Dissolve 0.100 g of NADH (FW = 709.4, product number N 8129, Sigma, St. Louis, Mo., ≈98 percent) in approximately 40 mL of DI water contained in a 50-mL volumetric flask. Dilute the resulting solution to the mark with DI water and mix it well. Use a digital pipet to transfer 1-mL aliquots of stock NADH reagent into 1.7-mL snap-cap vials (VWR, Cat. No. 20170-650) and store them in a freezer at -20°C where NADH thus prepared is stable for 6 weeks.

6.1.6 *NADH working solution.*—Remove one vial of stock NADH from the freezer and allow it to thaw at ambient temperature (about 20 minutes is required) while AtNaR2 reconstitutes. Then quantitatively transfer the stock NADH solution into in a 20-mL DA reagent tube as follows:

- Carefully pour the thawed NADH concentrate into the working reagent tube.

- Use a digital pipet to dispense 1,000 µL of phosphate buffer into the empty snap-cap vial.

- Recap the vial and invert it several times.

- Before flipping the cap up, tap it sharply with your finger to dislodge adherent droplets.

- Use a digital pipet equipped with a 10-mL liquid end to dispense 8.0 mL of phosphate buffer into the reagent tube and mix the contents well. This 10-mL volume of working NADH reagent, which is sufficient for 330 assays, is stable at 2°C to 8°C for at least 24 hours.

NOTE: The NWQL has found it convenient to use reagent kits (NECi product number DA-ARK-1) that contain a vial of freeze-dried AtNaR2 (3 units), an ampoule of reconstitution buffer, and a vial of freeze-dried NADH (2 mg). Use the procedure for frozen NADH concentrate described in 6.1.6 to prepare freeze-dried NADH.

6.2 *Colorimetric reagents*

6.2.1 *Sulfanilamide reagent (SAN).*—Slowly add 150 mL concentrated hydrochloric acid (HCl, ≈ 12*M*) to about 250 mL deionized water contained in a 500-mL volumetric flask. While the solution is still warm, add 5.0 g sulfanilamide ($C_6H_8N_2O_2S$, FW = 172.2) to the flask. Swirl the flask gently to dissolve the SAN. Dilute this reagent to the mark with deionized water and mix it well. Store SAN at room temperature in a clear glass or translucent plastic 500-mL bottle where it is stable for 6 months.

6.2.2 *N-(1-Naphthyl)ethylenediamine reagent (NED).*—Dissolve 0.5 g NED ($C_{12}H_{14}N_2$•2HCl, FW = 259.2) in about 400 mL of DI water contained in a 500-mL volumetric flask. Dilute this reagent to the mark with DI water and mix it well. Store NED at room temperature in an amber, 500-mL glass bottle where it is stable for 6 months.

7. Calibrants and Quality-Control Solutions

7.1 Use digital pipets and Class-A volumetric flasks to prepare secondary (Stock II) calibrants from a commercially obtained, certified 1,000 mg-N/L primary nitrate calibrant as indicated in *table 3*. In its current configuration, the DA prepares working calibrants for these assays by serial dilution of Stock II calibrants as indicated in *tables 4* and *5*. It is possible to calibrate the DA with individual, manually prepared working calibrants, but typically there is no advantage to this labor-intensive practice. Prepare Stock II calibrants monthly, transfer them to screw-cap, glass media bottles, and store them at 4°C. Follow vendor-specified storage temperatures and shelf lives for primary calibrants.

7.2 *Laboratory control samples and spike solutions*

7.2.1 The purpose of third-party-check (TPC) samples is to confirm and document the accuracy of instrument calibration. It is necessary, therefore, to prepare them from certified nitrate and nitrite solutions different from those used to prepare calibrants. Use digital pipets, Class-A volumetric flasks, and second-source, certified nitrate and nitrite solutions to prepare working TPC samples for standard- and low-level nitrate assays as indicated in *tables 6* and *8*, respectively. *Tables 7* and *9* provide preparation guidelines for nitrate-only TPCs. Nominal concentrations of low-level assay TPC samples in *tables 8* and *9* are slightly different from those used during validation experiments to simplify their preparation and to make their concentrations proportional (5-times less) to standard-level assay TPCs. Prepare TPCs monthly, transfer them to screw-cap glass media bottles, and store them at 4°C.

Table 3. Nitrate Stock II calibrants used for automated discrete analyzer calibration.

[ID, identifier; μg-N/μL, microgram nitrogen per microliter; μL, microliter; mL, milliliter; mg-N/L, milligram nitrogen per liter; NO_3^--N, nitrate nitrogen]

Stock II calibrant ID	Analyte	Stock I concentration μg-N/μL	Stock I dispensed volume (μL)	Stock II volume (mL)	Nominal concentration (mg-N/L)
S6	NO_3^--N (low-level)	1.0	400	100	4.00
S7	NO_3^--N (low-level)	1.0	60	100	0.60
S8	NO_3^--N	1.0	2,000	100	20.00
S9	NO_3^--N	1.0	300	100	3.00

Table 4. Standard-level concentration range nitrate working calibrants used for automated discrete analyzer (DA) calibration.

[ID, identifier; mg NO_3^--N/L, milligram nitrate nitrogen per liter]

Stock II calibrant ID	DA dilution factor notation	Dilution factor	Nominal concentration (mg NO_3^--N/L)
S8	S8: 1+3	4	5.00
S8	S8: 1+4	5	4.00
S8	S8: 1+5.5	6.5	3.08
S8	S8: 1+9	10	2.00
S8	S8: 1+19	20	1.00
S8	S8: 1+39	40	0.50
S9	S9: 1+11	12	0.25
S9	S9: 1+74	75	0.04

Table 5. Low-level concentration range nitrate working calibrants used for automated discrete analyzer (DA) calibration.

[ID, identifier; mg NO_3^--N/L, milligram nitrate nitrogen per liter]

Stock II calibrant ID	DA dilution factor notation	Dilution factor	Nominal concentration (mg NO_3^--N/L)
S6	S6: 1+3	4	1.000
S6	S6: 1+4	5	0.800
S6	S6: 1+5.5	6.5	0.615
S6	S6: 1+9	10	0.400
S6	S6: 1+19	20	0.200
S6	S6: 1+39	40	0.100
S7	S7: 1+11	12	0.050
S7	S7: 1+59	60	0.010

Table 6. Second-source, certified nitrate and nitrite solution volumes needed to prepare standard-level-assay third-party-check (TPC) samples in 100-mL quantities.

[mL, milliliter; ID, identifier; μL, microliter; mg-N/L, milligram nitrogen per liter; NO_3^-, nitrate; NO_2^-, nitrite]

Kone TPC ID	Second-source 1,000-mg NO_3^--N/L calibrant (μL)	Second-source 100-mg NO_2^--N/L calibrant* (μL)	Nominal concentration (mg-N/L)		
			NO_3^--N	NO_2^--N	NO_3^-+NO_2^--N
TPC_Low	48	20	0.48	0.02	0.50
TPC_Med	192	80	1.92	0.08	2.00
TPC_High	384	160	3.84	0.16	4.00

*If a certified 100 mg-N/L solution is not available commercially, prepare it from certified 1,000 mg-N/L solution after 1+9 dilution with deionized water.

Table 7. Second-source, certified nitrate solution volumes needed to prepare nitrate-only, standard-level-assay third-party-check (TPC) samples in 100-mL quantities.

[mL, milliliter; ID, identifier; μL, microliter; mg-N/L, milligram nitrogen per liter]

Kone TPC ID	Second-source 1,000 mg-N/L calibrant (μL)	Nominal nitrate concentration (mg-N/L)
TPC_Low	50	0.500
TPC_Med	200	2.000
TPC_High	400	4.000

Table 8. Second-source, certified nitrate and nitrite solution volumes needed to prepare low-level-assay third-party-check (LLTPC) samples in 100-mL quantities.

[mL, milliliter; ID, identifier; μL, microliter; mg-N/L, milligram nitrogen per liter; NO_3^-, nitrate; NO_2^-, nitrite]

Kone TPC ID	Second-source 100-mg NO_3^--N/L calibrant* (μL)	Second-source 100-mg NO_2^--N/L calibrant* (μL)	Nominal concentration (mg-N/L)		
			NO_3^--N	NO_2^--N	NO_3^-+NO_2^--N
LLTPC_Low	80	20	0.08	0.020	0.100
LLTPC_Med	320	80	0.32	0.080	0.400
LLTPC_High	640	160	0.64	0.160	0.800

*If certified 100 mg-N/L nitrate and nitrite solutions are not available commercially, prepare them from certified 1,000 mg-N/L solutions after 1+9 dilution with deionized water.

Table 9. Second-source, certified nitrate solution volumes needed to prepare nitrate-only, low-level-assay third-party-check (LLTPC) samples in 100-mL quantities.

[mL, milliliter; ID, identifier; μL, microliter; mg-N/L, milligram nitrogen per liter]

Kone TPC ID	Second-source 100 mg-N/L calibrant* (μL)	Nominal nitrate concentration (mg-N/L)
LLTPC_Low	100	0.100
LLTPC_Med	400	0.400
LLTPC_High	800	0.800

*If a certified 100 mg-N/L solution is not available commercially, prepare it from certified 1,000 mg-N/L solution after 1+9 dilution with deionized water.

7.2.2 Continuing calibration verification (CCV) samples are used to confirm and document that instrument calibration is maintained within specified limits during the course of analyses. Nitrate solutions prepared at concentrations 50 to 75 percent of assays' upper calibration limits are suitable. Stock II calibrants S9 and S7 (table 3) are convenient to use as standard- and low-level CCVs, respectively. Prepare CCVs monthly from the same certified nitrate solution used to prepare calibrants, transfer them to screw-cap glass media bottles, and store them at 4°C.

7.2.3 *Spike solutions.*—Prepare spike solutions for standard- and low-level nitrate assays with digital pipets, Class-A volumetric flasks, and the same certified nitrate solution used to prepare calibrants. Separately dilute 2.5 mL (standard level) and 0.5 mL (low level) of certified 1,000 mg-N/L nitrate solution to 50 mL with DI water. Resulting standard- and low-level nitrate spike solutions contain 0.05 μg-N/μL and 0.01 μg-N/μL, respectively. To increase sample or DI water blank concentration by 0.5 mg-N/L or 0.1 mg-N/L, dispense 10 μL of the appropriate spike solution into the conical well of a standard

2-mL analyzer cup, add 990 μL of sample or blank to it, and mix. This procedure dilutes spiked solutions by 1 percent. Prepare spike solutions monthly, transfer them to screw-cap glass media bottles, and store them at 4°C.

8. Sample Preparation

The DA enzymatic nitrate + nitrite methods require analysts to rinse and fill analyzer cups or tubes with well-shaken samples, place them into appropriate racks, and load racks into the sampler compartment. No other manual sample preparation is required.

9. Instrument Performance

The DA used to validate enzymatic reduction standard- and low-level nitrate assays has a nominal analysis rate of 600 tests per hour. However, for multistep assays such as these—four reagent additions and a total incubation time of about 14 minutes—the analysis rate is substantially less (300 tests per hour, perhaps) and is further reduced by samples that require dilution and by incidents of failed quality-control (QC) samples. Standard- and low-level assay volumes (sample + reagents) are 122 μL and 142 μL, respectively. For comparison, analysis rates for a single-channel, third generation continuous-flow (CF) analyzer performing similar assays was 90 tests per hour and per test sample and reagent volumes exceeded those of DA assays by about five times (Patton and others, 2002). Based on the 2011 price for NECi DA-ARK-1 reagent kits ($75.00; see note at the end of section 6), the per-assay cost of AtNaR2 and NADH for standard- and low-level methods is about 25 cents.

10. Calibration

Calibration functions for standard- and low-level assays are linear with linear least squares fit (Draper and Smith, 1966) correlation coefficients (r^2) equal to or greater than 0.999 as shown in *figure 4*. Calibration functions take the form $y = a + bx$, where y is the reagent-blank-corrected absorbance at 540 nm, x is the nitrate + nitrite concentration in mg-N/L, and a and b are the y-intercept and slope parameters. If there is slight bend off at higher concentrations, a second-order polynomial least-squares calibration function in the form $y = a + bx + cx^2$ might provide a better fit.

11. Procedure and Data Evaluation

Except as noted in sections 6 and 7, procedures for standard- and low-level assays were as specified in NWQL SOP INCF0452.2 (Gupta and others, 2011). *Table 10* identifies NWQL standard operating procedures (SOPs) that provide complete procedural details of USGS CFA-CdR methods against which we validated soluble DA-AtNaR2-reduction nitrate methods.

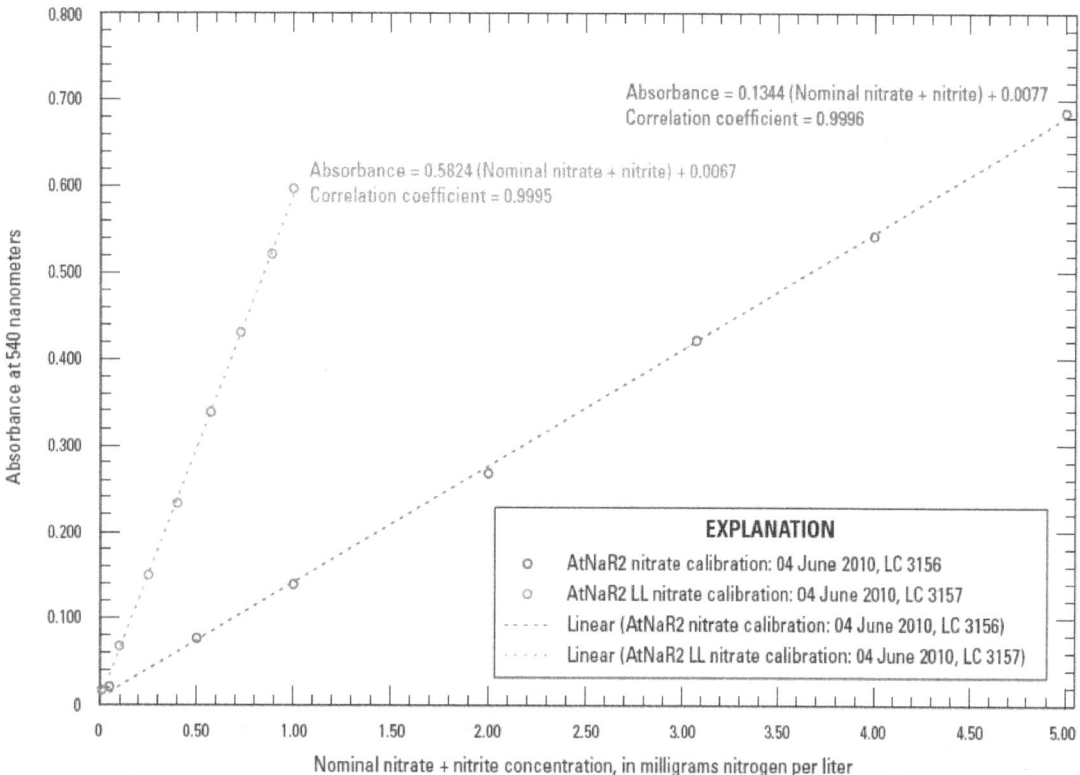

Figure 4. Typical calibration graphs for standard- and low-level concentration range enzymatic-reduction nitrate determination methods by automated discrete analyzer. (LL, low level; LC, National Water Quality Laboratory laboratory code)

Table 10. U.S. Geological Survey National Water Quality Laboratory (NWQL) laboratory codes and standard operating procedure (SOP) numbers for cadmium-reduction (CdR) and enzymatic-reduction (AtNaR2) nitrate determination methods referenced in this report.

Method name	Laboratory code	NWQL SOP number
Nitrate + nitrite, CdR, automated continuous-flow	1975	ID0163.6-1
Nitrate + nitrite, CdR, automated continuous-flow, low-level	1979	ID0200.2-1
Nitrate + nitrite, AtNaR2-reduction, automated discrete analyzer, standard-level concentration range	3156	INCF0452.2
Nitrate + nitrite, AtNaR2-reduction, automated discrete analyzer, low-level concentration range	3157	INCF0452.2

12. Calculations

12.1 We used vendor supplied software to acquire and process data from CFA (fASPac™ version 3.3, Astoria-Pacific, Clackamas, Oreg.) and DA (Aquakem 600™ versions 6.5, 7.0, and 7.2, Thermo Fisher Diagnostics, Fremont, Calif.) instrument platforms and to convert them into concentration units. Calibration functions for cadmium-reduction CFA methods were quadratic, linear least-squares fits (Draper and Smith, 1966) of the form $y = a + bx + cx^2$ (see section 10). Calibration functions for AtNaR2-reduction DA methods were typically linear least-squares fits.

12.2 We used Microsoft Office 2003 Excel™ to compile data acquired from instrument-specific software packages, to perform arithmetic and linear least-squares regression parameter calculations, and to prepare most graphical representations of data in this report. We used Microcal Origin Pro 8.0™ to perform Kolmogorov-Smirnov statistical tests of normality on spike recovery datasets and t-tests and Wilcoxon signed-rank tests on populations on paired CFA-CdR and DA-AtNaR2 nitrate + nitrite concentration data for surface water and groundwater.

12.3 Software packages identified in Section 12.1 provide for automatic application of dilution factors—the number by which a measured concentration must be multiplied to obtain the analyte concentration in the sample prior to dilution. Automatic, online dilution was not possible with the CFA equipment we used for work reported here, so we diluted off-scale samples manually using electronic pipets. The DA software also provides entry fields for offline dilution factors, but because this instrument was capable of up to 120-fold online sample dilution, manual offline sample dilution was rarely necessary. When both offline and online dilution factors are associated with the same sample, the dilution factor applied is the product of the two. Although dilution factors are applied identically by CFA and DA software applications, the factors are entered differently. CFA software requires entering the sum of 1 part sample + n parts diluent. For example, entering values of 2, 5, and 10 into the CFA software dilution factors fields indicate sample-to-diluent proportions of 1+1, 1+4, and 1+9—that is, two-, five-, and tenfold dilutions. DA software requires entry only of the parts of diluent added to 1 part of sample, and dilution factor entry fields always appear as 1+n. Therefore, entering values of 1, 4, and 9 into DA software dilution factor fields result in two-, five-, and tenfold dilutions.

12.4 Control limits for TPC solutions were calculated according to protocols developed by the USGS Branch of Quality Systems (BQS) for use with their Inorganic Blind Sample Program (IBSP). BQS control limit estimates use a robust (median-based) statistic equivalent to mean-based standard deviation, which BQS denotes f_σ. BQS suggests three methods for estimating f_σ.

12.4.1 Use regression equations of analyte concentration in relation to f_σ, which are tabulated on the BQS Web site (*http://bqs.usgs.gov/ibsp/*) by year.

12.4.2 f_σ = 75 percent of LT-MDL or provisional MDL (set at 0.04 mg-N/L and 0.008 mg-N/L for standard- and low-level DA-AtNaR2 assays during the first year of operation following Office of Water Quality approval), or

12.4.3 f_σ = 5 percent of each nominal TPC concentration.

The BQS suggests estimating f_σ by all three methods and selecting the largest of the three for each nominal concentration. By NWQL Nutrients Unit convention, upper and lower control limits (UCL and LCL) are set at 1.5 times f_σ ± the nominal TPC concentration.

13. Reporting Results

13.1 Reporting units for nitrate + nitrite and nitrite concentrations are milligrams nitrogen per liter (mg-N/L) in accordance with longstanding USGS conventions. A table at the front of this report provides factors necessary to convert these units into several other commonly used concentration units.

13.2 We report concentrations such that the rightmost digit (called the least significant digit) represents the uncertainty in the analytical result (Novak, 1985; Hansen, 1991; U.S. Geological Survey, 2002). The least significant digit is determined using guidance outlined by the American Society for Testing and Materials (1999). Presently (2011), the NWQL reports results in the national database to the least significant digit plus one additional digit.

14. Detection Limits, Precision, Spike Recovery, and Bias

14.1 As listed in *table 11*, MDLs for standard- and low-level DA-AtNaR2 nitrate + nitrite assays were 0.02 mg-N/L and 0.002 mg-N/L, respectively, which we calculated in accordance with EPA guidelines (U.S. Environmental Protection Agency, 1997). By NWQL consensus, interim reporting limits (IRLs) for standard- and low-level DA-AtNaR2 nitrate + nitrite assays will be set at 0.04 mg-N/L and 0.008 mg-N/L, respectively, during the first year of routine operation.

14.2 *Table 12* lists the average, standard deviation, and other information related to TPC samples that we analyzed along with standard-level environmental water samples during DA-AtNaR2 assay validation work between November 30, 2007, and August 8, 2008. *Table 12* also contains the same parameters for low-level TPCs that we analyzed during a single day on February 15, 2007. Inspection of *table 12* reveals that over

Table 11. Data and calculations used to estimate method detection limits (MDL) for nitrate + nitrite determination with soluble AtNaR2 nitrate reductase by automated discrete analysis (DA).

[mg-N/L, milligram nitrogen per liter; NO$_3^-$, nitrate; NO$_2^-$, nitrite; SL, standard level; LL, low level; LT, long term]

| Target concentration SL (LL) | Nitrate + nitrite concentration (mg-NO$_3^-$-N + NO$_2^-$-N/L) | |
| | Concentration found | |
	AtNaR2 SL	AtNaR2 LL
0.08 (0.020)	0.0767	0.0190
0.08 (0.020)	0.0748	0.0189
0.08 (0.020)	0.0732	0.0180
0.08 (0.020)	0.0870	0.0186
0.08 (0.020)	0.0746	0.0187
0.08 (0.020)	0.0731	0.0180
0.08 (0.020)	0.0770	0.0201
0.08 (0.020)	0.0826	0.0181
Average	0.0774	0.0187
Standard deviation	0.0049	0.0007
Number of values	8	8
Degrees of freedom	7	7
t-value (1-sided, 99 percent)	2.998	2.998
MDL	0.02	0.002
2011 CFA-CdR (LT-MDL)	0.02	0.008

a period of about 9 months, standard-level assay TPC results at high, medium, and low concentrations agree within about 3 percent. Again with reference to *table 12*, within-day results for the low-level AtNaR2 nitrate assay at high, medium, and low TPC concentrations agree within about 1 percent. *Figure 5* provides a plot of standard-level assay TPC data obtained between November 30, 2007, and August 8, 2008, in control chart format. Upper and lower control limits in *figure 5* correspond to those listed in *table 12*, which are in accordance with IBSP guidance as described in section 12.4.

Table 13 provides within-day and between-day precision summaries of AtNaR2-reduction nitrate + nitrite assays performed on water samples between May and July of 2007. With reference to *table 13*, within-day precision was about ±1 percent but decreased as concentrations approached the detection limit. Between-day precision was remarkably good (on the order of ±5 percent) considering that on the second analysis date most samples were past their 30-day holding time limit. *Table 14* provides within-day precision summaries for low-level AtNaR2-reduction nitrate + nitrite assays. With reference to *table 14*, within-day precision was about ±3 percent but decreased as concentrations approached the detection limit.

Tables 15A–C and *16A–C* list recoveries of nitrate spiked into reagent water, surface water, and groundwater at about 5 times and 50 times standard- and low-level MDLs for standard- and low-level DA-AtNaR2 assays, respectively. With reference to these tables, recoveries typically were 100±20 percent, which are well within NWQL-specified criteria for accepted analytes

Table 12. Third-party-check (TPC) sample nitrate + nitrite data summary for automated discrete analysis with soluble AtNaR2 nitrate reductase.

[Standard-level (SL) data were collected between November 30, 2007, and August 8, 2008. Low-level (LL) data were collected on February 15, 2007. NO$_2^-$, nitrite; NO$_3^-$, nitrate; mg-N/L, milligram nitrogen per liter; ID, identifier; H, high; M, medium; L, low]

	SL NO$_2^-$ + NO$_3^-$ (mg-N/L)			LL NO$_2^-$ + NO$_3^-$ (mg-N/L)		
Reference sample ID	TPC-H	TPC-M	TPC-L	TPC-H	TPC-M	TPC-L
Most probable value	4.00	2.00	0.50	0.750	0.500	0.125
Upper control limit	4.33	2.17	0.54	0.816	0.544	0.136
Lower control limit	3.67	1.83	0.46	0.684	0.465	0.114
Average concentration	4.03	1.99	0.49	0.783	0.536	0.129
Standard deviation (SD)	0.11	0.06	0.02	0.005	0.003	0.002
Relative SD, percent	2.7	3.1	3.3	0.6	0.6	1.2
Number of points	143	143	143	10	10	10

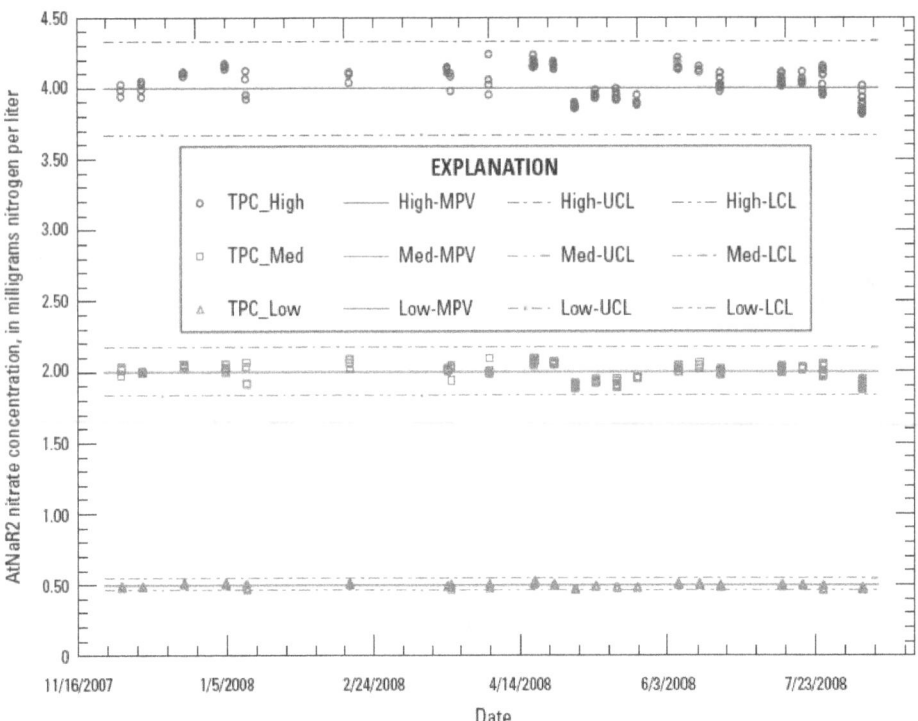

Figure 5. Control chart of third-party-check (TPC) sample results for standard-level automated discrete analyzer nitrate + nitrite methods using soluble AtNaR2 nitrate reductase. (MPV, most probable value; UCL, upper control limit; LCL, lower control limit)

(Green and Foreman, 2005, table 1). Each seven-replicate dataset used to estimate spike recoveries was normally distributed at the $p = 0.05$ probability level on the basis of Kolmogorov-Smirnov statistical tests of normality (Pollard, 1979). Predictably, recovery of nitrate spiked near the standard-level assay MDL concentration (see *table 15A*) was more variable. See section 7.2.3 for details of how to prepare the seven replicate spikes at each concentration level in the three matrices.

Statistical and graphical comparisons of paired nitrate results obtained with USGS-approved CFA-CdR and DA-AtNaR2 methods provided in the section "Analytical Performance and Comparative Results" that follows permit direct assessment of between-method bias.

Analytical Performance and Comparative Results

Background Information

During initial development of DA enzymatic reduction nitrate + nitrite assays, we were surprised to observe that analytical results obtained did not compare as well with CFA-CdR reference assay results as those obtained with CFA-enzymatic reduction assays that were developed previously (Patton and others, 2002). This was the case not only for nitrate reductase purified from corn seedlings (NaR1™), but also for two other

commercially available nitrate reductases—*Aspergillus* sp. NADPH:nitrate reductase (product number N 7625, Sigma, St. Louis, Mo.) and recombinant, bispecific NAD(P)H:nitrate reductase from *Pichia angusta* (YNaR1™, Nitrate Elimination Company, Lake Linden, Mich.). We eventually discovered that these performance issues had two distinct, temperature-dependent causes: (1) high-phenolic-content humic acid (HA) irreversibly inhibits these enzymes at reaction temperatures greater than 20°C, and (2) except for NaR1, temperatures above about 25°C decrease the activity of these enzymes. In the sections that follow, we present descriptions and results of experiments with these three enzymes and recombinant NADH:nitrate reductase from *Arabidopsis thaliana* (AtNaR2™, Nitrate Elimination Company, Lake Linden, Mich.) that demonstrate the latter's particular suitability as an analytical reagent for routine analysis of nitrate in environmental water samples on DA platforms.

Effects of Temperature and Dissolved Organic Matter on AtNaR2 Activity

Our initial characterization of AtNaR2 as an analytical reagent began with a replication of experiments we had performed earlier to elucidate the effects of reaction temperature and HA concentration on YNaR1 activity. A major finding of this research was that AtNaR2 activity remains high—sufficient for quantitative reduction of 5 mg NO_3^--N/L to nitrite in less than 10 minutes—at reaction temperatures ranging from 10°C to 37°C and at HA concentrations up to 20 mg/L. We similarly assessed the susceptibility of *Aspergillus* sp.

Table 13. Within-day and between-day replicate agreement for nitrate + nitrite analyses of surface-water samples and groundwater samples analyzed by the AtNaR2-reduction automated discrete analyzer method (NWQL laboratory code 3156).

[NWQL, National Water Quality Laboratory; ID, identifier; mg-N/L, milligram nitrogen per liter]

NWQL sample ID	First analysis date	Result-1	Result-2	Within-day difference (mg-N/L)	Within-day difference (percent)	Second analysis date	Days between analyses	Result-3	Between-day difference (mg-N/L)	Between-day difference (percent)
20081580131	6/6/2008	0.05	0.05	0.00	−0.7	7/25/2008	49	−0.02	0.07	149.3
20081560054	6/6/2008	0.06	0.05	0.01	8.8	7/25/2008	49	0.02	0.04	69.0
20081570105	6/6/2008	0.07	0.07	0.00	4.0	7/25/2008	49	0.03	0.04	54.5
20081340114	5/16/2008	0.35	0.35	0.00	0.2	7/25/2008	70	0.32	0.03	8.8
20081340115	5/16/2008	0.35	0.35	0.00	1.3	7/25/2008	70	0.22	0.13	36.5
20081580042	6/6/2008	0.49	0.49	0.00	−0.4	7/25/2008	49	0.47	0.02	4.7
20081570106	6/6/2008	0.58	0.59	0.00	−0.3	7/25/2008	49	0.55	0.03	5.3
20081370088	5/16/2008	1.19	1.20	−0.01	−1.0	7/25/2008	70	1.18	0.01	0.8
20081580126	6/6/2008	1.36	1.35	0.00	0.3	7/25/2008	49	1.28	0.07	5.3
20081970033	7/18/2008	1.71	1.70	0.01	0.4	7/25/2008	7	1.72	−0.01	−0.7
20081970032	7/18/2008	1.71	1.71	0.00	0.1	7/25/2008	7	1.73	−0.02	−1.1
20081480025	6/6/2008	2.56	2.52	0.04	1.4	7/25/2008	49	1.68	0.88	34.4
20081580041	6/6/2008	2.67	2.68	0.00	−0.1	7/25/2008	49	2.61	0.06	2.3
20081350062	5/16/2008	2.83	2.86	−0.03	−0.9	7/25/2008	70	2.85	−0.02	−0.8
20081370089	5/16/2008	3.08	3.07	0.00	0.1	7/25/2008	70	3.00	0.07	2.4
20081580124	6/6/2008	3.14	3.10	0.04	1.3	7/25/2008	49	3.05	0.09	2.9

Table 14. Between-day replicate agreement for nitrate + nitrite analyses of surface-water samples and groundwater samples by the low-level AtNaR2-reduction automated discrete analyzer method. Units for cadmium reduction (CdR) and nitrate reductase (AtNaR2) reduction methods are milligrams nitrate + nitrite nitrogen per liter.

[NWQL, National Water Quality Laboratory; ID, identifier; LC, NWQL laboratory code; SD, standard deviation; RSD, percent relative standard deviation; MC, U.S. Geological Survey National Water Information System sample medium code; WS, surface-water medium code; --, no data; WG, groundwater medium code]

NWQL sample ID	CdR (LC 1979)	AtNaR2 trial 1	AtNaR2 trial 2	AtNaR2 trial 3	AtNaR2 average	SD	RSD	MC
20060820056-2	2.792	2.779	2.758	2.907	2.815	0.081	2.9	WS
20060820058-2	1.949	1.951	1.931	2.024	1.969	0.049	2.5	WS
20060820080-4	1.299	1.293	1.277	1.308	1.293	0.015	1.2	WS
20060820082-4	1.042	1.036	1.040	1.058	1.045	0.012	1.1	WS
20060820077-4	0.652	0.657	0.681	0.667	0.668	0.012	1.9	WS
20060860110-2	0.171	0.193	0.186	--	0.190	0.005	2.8	WS
20060830065-2	0.133	0.135	0.129	0.133	0.132	0.003	2.3	WS
20060820079-4	0.120	0.122	0.110	0.116	0.116	0.006	5.1	WS
20060820078-4	0.089	0.097	0.093	0.094	0.095	0.002	2.1	WS
20060820075-4	0.083	0.086	0.087	0.087	0.087	0.001	0.8	WS
20060820074-4	0.080	0.080	0.077	0.080	0.079	0.002	2.3	WS
20060860113-5	0.053	0.060	0.059	--	0.060	0.000	0.7	WS
20060860111-2	0.040	0.043	0.043	--	0.043	0.000	0.2	WS
20060830066-2	0.023	0.020	0.027	0.024	0.024	0.004	15.6	WS
20060860108-2	0.013	0.016	0.010	--	0.013	0.004	34.9	WS
20060830067-2	0.007	0.007	0.012	0.009	0.009	0.003	28.0	WS
20060880004-4	1.078	1.074	1.095	--	1.084	0.015	1.3	WG
20060870028-2	0.241	0.263	0.252	--	0.258	0.008	2.9	WG
20060880003-4	0.002	0.004	0.004	--	0.004	0.000	5.1	WG

Table 15. Recovery of replicate spikes in reagent-water, surface-water, and groundwater matrices by standard-level (SL) AtNaR2-reduction nitrate + nitrite assay (NWQL laboratory code 3156).

[NWQL, National Water Quality Laboratory; n = 7 unless otherwise indicated; spike composition = 10 microliter (µL) spike solution + 990 µL sample; NO_3^--N, nitrate nitrogen; ID, identifier; MC, U.S. Geological Survey medium code; LL CdR, NWQL laboratory code 1979; mg-N/L, milligram nitrogen per liter; CdR, NWQL laboratory code 1975; DI, deionized; NA, not applicable; --, no data; <, less than; ±, plus or minus; WS, surface water; WG, groundwater]

A.   Recovery of replicate 0.02 mg NO_3^--N/L spikes.

NWQL laboratory ID	MC	Approved cadmium-reduction (CdR) methods		Standard-level AtNaR2 nitrate-reductase reduction method			
		LL CdR (mg-N/L)	CdR (mg-N/L)	Unspiked (mg-N/L)	Spike added (mg-N/L)	Found (mg-N/L)	Recovery (percent)
DI water	NA	--	--	< 0.02	0.02	0.01 ± 0.03	59
20110700114	WS	0.100	--	[1]0.09 ± 0.00	0.02	0.11 ± 0.00	97
20110760169	WS	0.519	--	0.52 ± 0.00	0.02	0.54 ± 0.01	114
20110770085	WS	[2]2.32	--	2.15 ± 0.03	0.02	2.16 ± 0.03	32
20110750033	WG	--	0.10	[1]0.10 ± 0.00	0.02	0.11 ± 0.00	103
20110750036	WG	--	0.56	0.58 ± 0.00	0.02	0.60 ± 0.01	123
20110770052	WG	--	2.61	2.54 ± 0.02	0.02	2.52 ± 0.02	−55

[1]n = 6. [2]Diluted 1+4.

B.   Recovery of replicate 0.10 mg NO_3^--N/L spikes.

NWQL laboratory ID	MC	Approved cadmium-reduction (CdR) methods		Standard-level AtNaR2 nitrate-reductase reduction method			
		LL CdR (mg-N/L)	CdR (mg-N/L)	Unspiked (mg-N/L)	Spike added (mg-N/L)	Found (mg-N/L)	Recovery (percent)
DI water	NA	--	--	< 0.02	0.10	0.10 ± 0.005	98
20110700114	WS	0.100	--	[1]0.09 ± 0.00	0.10	0.20 ± 0.00	113
20110760169	WS	0.519	--	0.52 ± 0.00	0.10	0.62 ± 0.01	107
20110770085	WS	[2]2.32	--	2.15 ± 0.03	0.10	2.26 ± 0.03	109
20110750033	WG	--	0.10	[1]0.10 ± 0.00	0.10	0.21 ± 0.00	110
20110750036	WG	--	0.56	0.58 ± 0.00	0.10	0.69 ± 0.01	111
20110770052	WG	--	2.61	2.54 ± 0.02	0.10	2.62 ± 0.03	82

[1]n = 6. [2]Diluted 1+4.

C.   Recovery of replicate 0.50 mg NO_3^--N/L spikes.

NWQL laboratory ID	MC	Approved cadmium-reduction (CdR) methods		Standard-level AtNaR2 nitrate-reductase reduction method			
		LL CdR (mg-N/L)	CdR (mg-N/L)	Unspiked (mg-N/L)	Spike added (mg-N/L)	Found (mg-N/L)	Recovery (percent)
DI water	NA	--	--	< 0.02	0.50	0.51 ± 0.01	101.7
20110700114	WS	0.100	--	[1]0.09 ± 0.00	0.50	[1]0.59 ± 0.03	100.0
20110760169	WS	0.519	--	0.52 ± 0.00	0.50	[1]1.06 ± 0.05	108.0
20110770085	WS	[2]2.32	--	2.15 ± 0.03	0.50	2.68 ± 0.05	105.0
20110750033	WG	--	0.10	[1]0.10 ± 0.00	0.50	0.63 ± 0.01	106.0
20110750036	WG	--	0.56	0.58 ± 0.00	0.50	1.12 ± 0.01	107.0
20110770052	WG	--	2.61	2.54 ± 0.02	0.50	3.03 ± 0.03	98.9

[1]n = 6. [2]Diluted 1+4.

Table 16. Recovery of replicate spikes in reagent-water, surface-water, and groundwater matrices by low-level (LL) AtNaR2-reduction nitrate + nitrite assay (NWQL laboratory code 3157).

[NWQL, National Water Quality Laboratory; n = 7 unless otherwise indicated; spike composition = 10 microliter (µL) spike solution + 990 µL sample; NO_3^--N, nitrate nitrogen; ID, identifier; MC, U.S. Geological Survey medium code; LL CdR, NWQL laboratory code 1979; mg-N/L, milligram nitrogen per liter; CdR, NWQL laboratory code 1975; DI, deionized; NA, not applicable; --, no data; ±, plus or minus; WS, surface water; WG, groundwater]

A. Recovery of replicate 0.02 mg NO_3^--N/L spikes.

NWQL laboratory ID	MC	Approved cadmium-reduction (CdR) methods		Low-level AtNaR2 nitrate-reductase reduction methods			
		LL CdR (mg-N/L)	CdR (mg-N/L)	Unspiked (mg-N/L)	Spike added (mg-N/L)	Found (mg-N/L)	Recovery (percent)
DI water	NA	--	--	0.002 ± 0.000	0.020	0.021 ± 0.003	95.0
20110700114	WS	0.100	--	0.100 ± 0.001	0.020	0.121 ± 0.002	104.0
20110760169	WS	0.519	--	0.517 ± 0.008	0.020	0.533 ± 0.003	79.4
20110770085	WS	[1]2.32	--	off-scale	--	--	--
20110750033	WG	--	0.10	0.101 ± 0.001	0.020	0.125 ± 0.006	119
20110750036	WG	--	0.56	0.606 ± 0.005	0.020	0.626 ± 0.011	100
20110770052	WG	--	2.61	off-scale	--	--	--

[1]Diluted 1+4.

B. Recovery of replicate 0.10 mg NO_3^--N/L spikes.

NWQL laboratory ID	MC	Approved cadmium-reduction (CdR) methods		Low-level AtNaR2 nitrate-reductase reduction methods			
		LL CdR (mg-N/L)	CdR (mg-N/L)	Unspiked (mg-N/L)	Spike added (mg-N/L)	Found (mg-N/L)	Recovery (percent)
DI water	NA	--	--	0.002 ± 0.000	0.100	0.102 ± 0.006	100.0
20110700114	WS	0.100	--	0.100 ± 0.001	0.100	0.202 ± 0.002	101.0
20110760169	WS	0.519	--	0.517 ± 0.008	0.100	0.615 ± 0.004	98.0
20110770085	WS	[1]2.32	--	off-scale	--	--	--
20110750033	WG	--	0.10	0.101 ± 0.001	0.100	0.217 ± 0.004	116.0
20110750036	WG	--	0.56	0.606 ± 0.005	0.100	0.713 ± 0.006	106
20110770052	WG	--	2.61	off-scale	--	--	--

[1]Diluted 1+4.

C. Recovery of replicate 0.50 mg NO_3^--N/L spikes.

NWQL laboratory ID	MC	Approved cadmium-reduction (CdR) methods		Low-level AtNaR2 nitrate-reductase reduction methods			
		LL CdR (mg-N/L)	CdR (mg-N/L)	Unspiked (mg-N/L)	Spike added (mg-N/L)	Found (mg-N/L)	Recovery (percent)
DI water	NA	--	--	0.002 ± 0.000	0.500	0.499 ± 0.006	99.0
20110700114	WS	0.100	--	0.100 + 0.001	0.500	[1]0.58 ± 0.03	97.0
20110760169	WS	0.519	--	0.517 ± 0.008	0.500	off-scale	--
20110770085	WS	[2]2.32	--	off-scale	--	--	--
20110750033	WG	--	0.10	0.101 ± 0.001	0.500	0.65 ± 0.01	110.0
20110750036	WG	--	0.56	0.606 ± 0.005	0.500	off-scale	--
20110770052	WG	--	2.61	off-scale	--	--	--

[1]n = 6. [2]Diluted 1+4.

NADPH:nitrate reductase to HA inhibition in relation to reaction temperature.

Figure 6 provides a graphical summary of results from these kinetics experiments. Points plotted in *figure 6* are proportional to nitrite concentrations recorded after about 9 minutes of enzymatic reaction time. In *figure 6*, nitrite concentrations determined in 5 mg-N/L nitrate solutions spiked with 20 mg/L HA are plotted with hollow symbols; those for 5 mg-N/L nitrate solutions not containing HA are plotted with solid symbols. To enhance clarity in this figure, only points from the highest temperature tested in nitrate solutions not containing HA are shown. With reference to *figure 6*, apparent activity of AtNaR2 was little affected by 20 mg/L HA at reaction temperatures ranging between 10°C and 37°C. However, apparent activities of the other three nitrate reductases tested began to decrease precipitously when reaction temperatures exceeded 20°C. As expected from prior experiments, apparent activity of AtNaR1 and YNaR1 were comparable at 10°C. Furthermore, at 37°C for nitrate solutions not containing HA, apparent activities of YNaR1 and *Aspergillus* sp. nitrate reductases were substantially less than that of nitrate reductase purified from corn, which approached that of AtNaR2. In summary, apparent activities of the four nitrate reductases

tested in solutions containing 5 mg NO_3^--N/L and 20 mg HA/L at the DA reaction zone temperature of 37°C were as follows: AtNaR2 >> NaR1 > NADPH:NaR ≈ YNaR1.

Data shown in *figure 7* demonstrate that AtNaR2 and NADH concentrations equal to those optimized in previous work with NaR1 and YNaR1 were sufficient to reduce nitrate to nitrite quantitatively within 10 minutes at 37°C for nitrate concentrations in the range of 0.05 to 5.0 mg-N/L. Nitrate solutions used in these experiments did not contain HA.

We next confirmed that YNaR1 and AtNaR2 reactivity at 37°C for nitrate solutions spiked with HA were the same on the DA platform as on the CFA kinetics platform. For this work, we prepared three series of nitrate standards in deionized water—five each at concentrations of 0.25 mg-N/L, 2.5 mg-N/L, and 5.0 mg-N/L—and amended each concentration series with 0, 5, 10, 15, and 20 mg/L HA. We then analyzed these solutions successively on the DA, first with YNaR1 and then with AtNaR2 enzyme reagent. *Figure 8* provides graphical summaries of these experimental results. As expected, apparent nitrate concentrations for test solutions decreased as HA concentration increased with YNaR1 reagent but remained high and constant with AtNaR2 reagent.

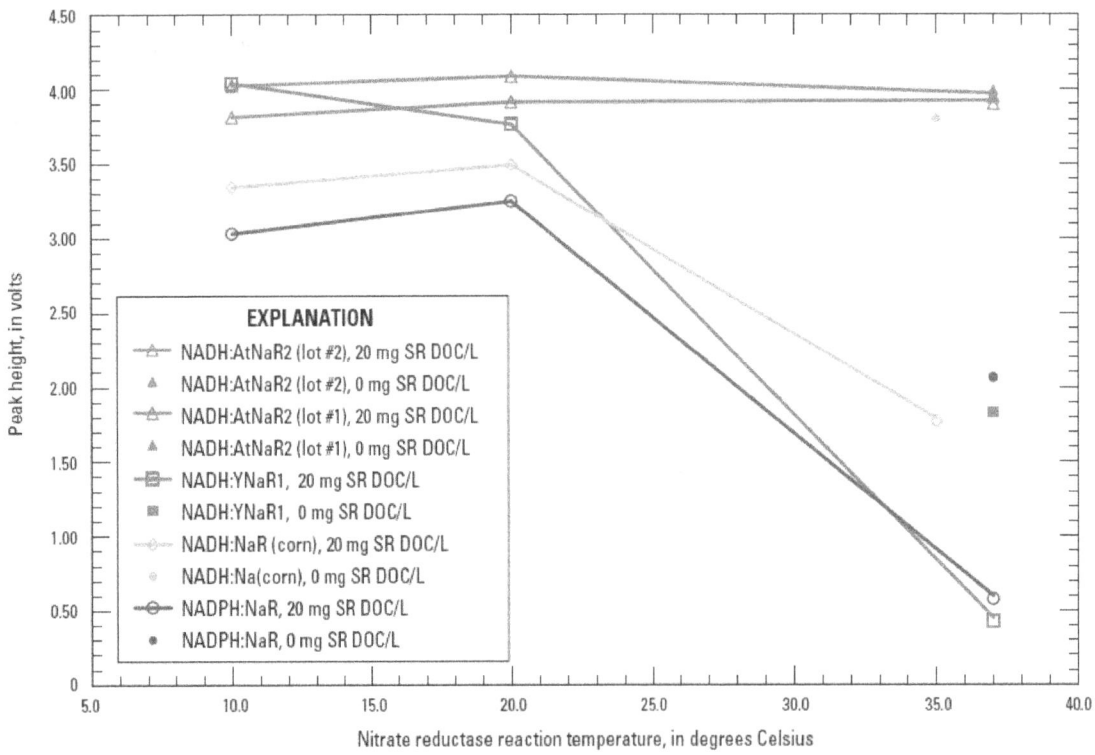

Figure 6.   Activity of different nitrate reductases in relation to reaction temperature for 5 milligrams nitrogen per liter nitrate solutions with and without added Suwannee River (SR) humic acid (SR DOC in explanation). NADPH:NaR is from *Aspergillus* species. Data were collected from late August through November 2005. (mg, milligram; L, liter; NADH, nicotinamide adenine dinucleotide in reduced form; NADPH, nicotinamide adenine dinucleotide phosphate in reduced form)

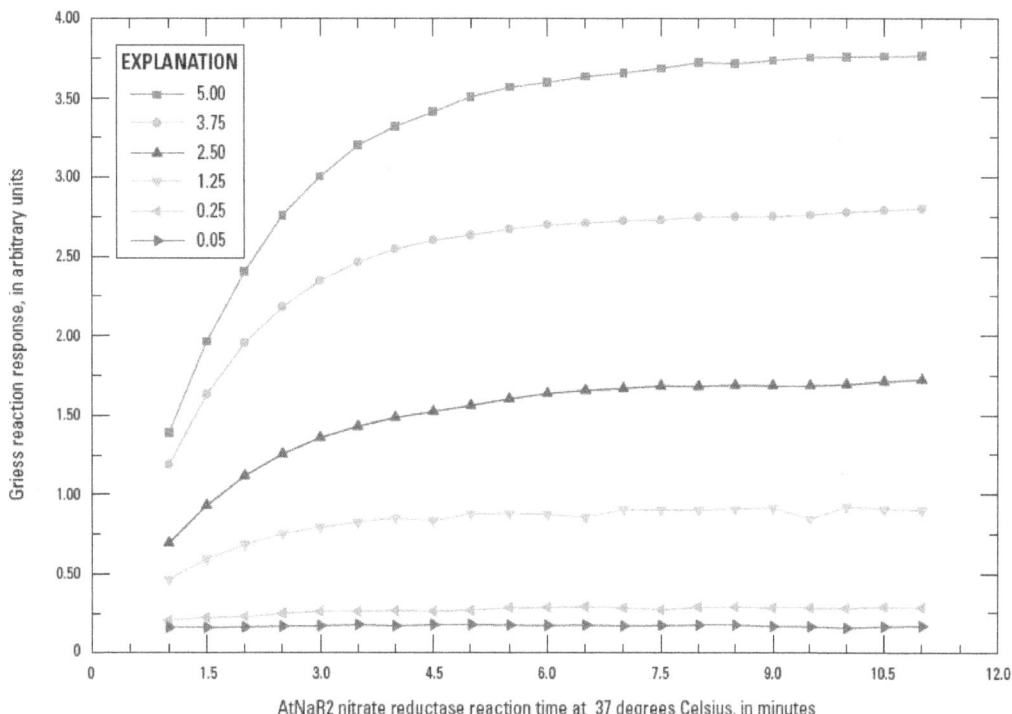

Figure 7. Nitrate to nitrite reduction rates by AtNaR2 nitrate reductase at 37 degrees Celsius for nitrate concentrations of 0.05 to 5.00 milligrams nitrogen per liter for each experiment.

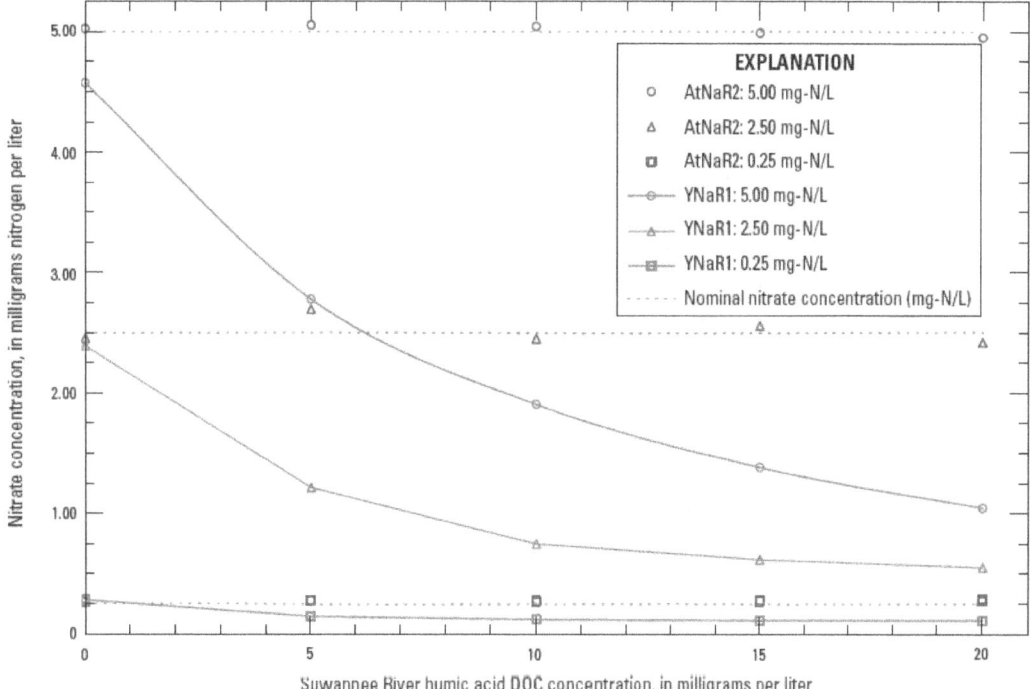

Figure 8. Nitrate recoveries from deionized water solutions containing 0.25 milligrams nitrogen per liter (mg-N/L), 2.50 mg-N/L, and 5.00 mg-N/L and incrementally increasing Suwannee River humic acid concentrations for automated discrete analyzer nitrate assays that differed only in the nitrate reductase (YNaR1 or AtNaR2) used to reduce nitrate to nitrite. Humic acid concentrations are expressed as milligrams dissolved organic carbon (DOC) per liter. Nominal temperature was 37 degrees Celsius for the enzymatic reduction reaction and the Griess reagent indicator reaction. Reaction time was 10 minutes for the enzymatic reduction step.

Reagent Stability

Experimental results summarized in *figure 9* established that the useful lifetime for working AtNaR2 reagent is about 18 hours in the 4°C environment of the DA's reagent compartment. We found, however, that preparing working AtNaR2 reagent in 0.05 *M*, pH 7.5 MOPS buffer increased its useful lifetime at 4°C to about 3 days (see *fig. 9*). Despite the longer AtNaR2 working reagent storage life afforded by MOPS buffer, we opted to continue using phosphate buffer to maintain continuity with our previous work. Addition of 25 percent glycerol to either phosphate or MOPS assay buffers stabilized working AtNaR2 reagent somewhat, but it also reduced assay sensitivity by 10–15 percent. With final reference to *figure 9*, NADH reagents prepared in either phosphate or MOPS buffers remained stable at 4°C for the 4-day duration of these experiments.

Interference by Anionic and Cationic Sample Matrix Constituents

As shown in *figure 10*, chloride, bromide, and sulfate at up to 100 times NWQL median concentrations—1,515 mg/L, 15 mg/L, and 2,287 mg/L, respectively (see *table 17*)—had negligible effect on recovery of 2.5 mg NO_3^--N/L in relation to the anticipated NWQL interim reporting limit for the standard-level AtNaR2 assay of 0.04 mg-N/L. In *figure 10*, error bars indicate the standard deviation for three replicate nitrate determinations in each anion-amended test solution. The leftmost column, labeled DI, indicates the average concentration measured for 2.5 mg NO_3^--N/L in DI water that was not amended with anions. Median perchlorate concentrations were not available, and those selected coincide with 10 percent, 50 percent, and 100 percent of full-scale nitrate concentrations in the standard-level DA-AtNaR2 assay.

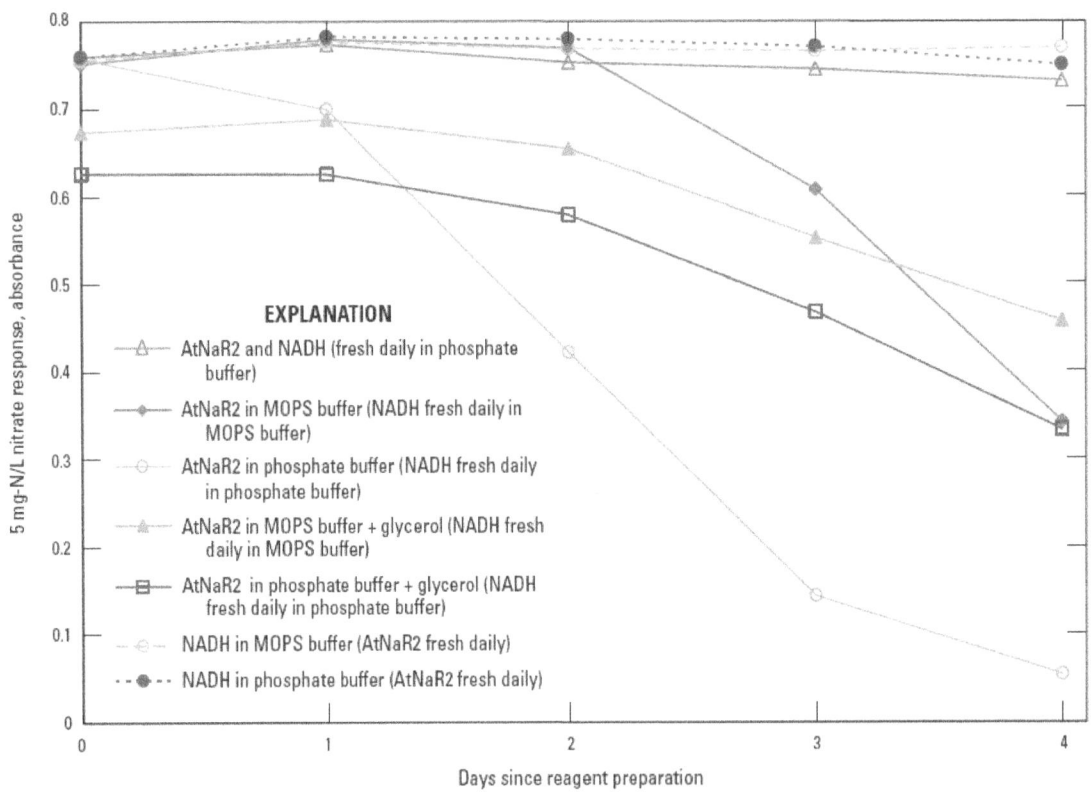

Figure 9.   AtNaR2 storage stability plots. In-use and between-day storage for all reagent solutions was at 4 degrees Celsius. (mg-N/L, milligram nitrogen per liter; NADH, nicotinamide adenine dinucleotide in reduced form; MOPS, 3-N-morpholino-propansulfonic acid)

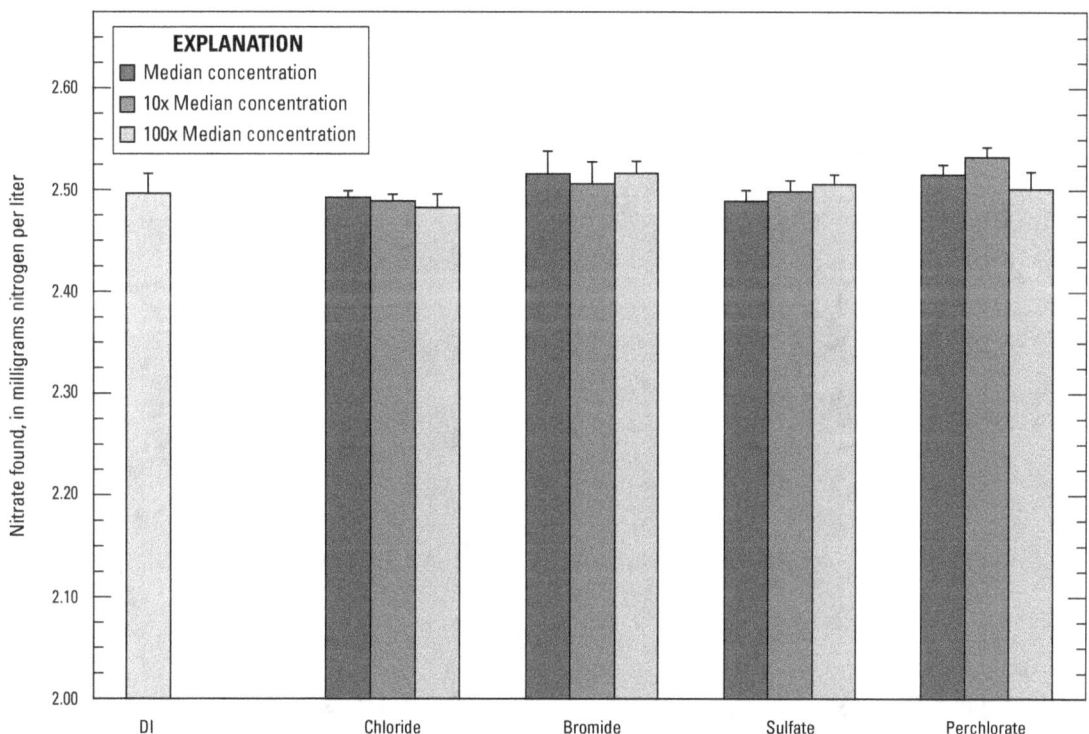

Figure 10.   Standard-level discrete analyzer AtNaR2 assay recovery of 2.50 milligrams nitrogen per liter (mg-N/L) nitrate from deionized (DI) water amended with various anions at 1, 10, and 100 times their National Water Quality Laboratory annual median concentrations as listed in table 17. The leftmost column, labeled DI, indicates recovery of 2.50 mg-N/L nitrate from DI water not amended with anions. Column heights and associated error bars at tops of columns indicate the average and standard deviation of nitrate recovery for three replicate determinations of each test solution.

Table 17.   Nominal concentrations of anions tested for possible interference in the discrete analyzer AtNaR2 nitrate assay.

[FW, formula weight; mg/L, milligram per liter]

Constituent	FW	Concentration (mg/L)		
		Median	10x Median	100x Median
Chloride	35.45	15.15	151.5	1,515
Bromide	79.90	0.15	1.5	15
Sulfate	96.06	22.87	228.7	2,287
Perchlorate*	99.45	0.5	2.5	5.0

*Perchlorate concentrations coincide with 10 percent, 50 percent, and 100 percent of full-scale nitrate concentrations for the standard-level nitrate assay.

Companion *figure 11* provides a graphical summary of the effects of metal ions at concentrations equal to the NWQL median, 10 times the median, and 100 times the median (see *table 18*) on recovery of 2.5 mg-N/L nitrate solutions. Error bars in *figure 11* indicate the standard deviation of three replicate nitrate determinations by the DA-AtNaR2 nitrate assay in each metal-ion-amended test solution. The leftmost column, labeled DI, indicates the average concentration measured for 2.5 mg-N/L nitrate in DI water that was not amended with

metal ions. With the exception of nitrate recovery at a calcium concentration 100 times greater than the NWQL median, other cations tested had only minor effects (less than ±2 percent) on nitrate recovery in relation to the metal-free 2.5 mg-N/L nitrate test solution. Additional experiments suggested that low nitrate recovery (about 85 percent) from calcium test solutions at concentrations 100 times the median (prepared with calcium chloride) resulted from AtNaR2 inhibition by chloride counter ions (≈ 6,400 mg/L).

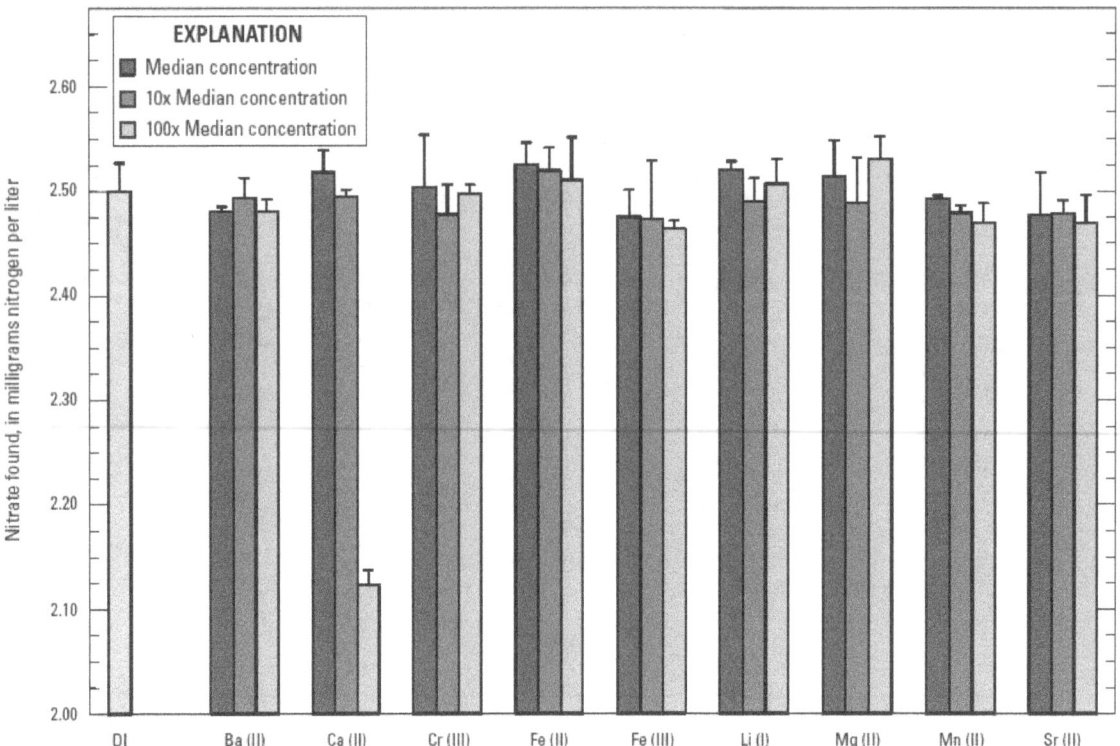

Figure 11. Standard-level discrete analyzer AtNaR2 assay recovery of 2.50 milligrams nitrogen per liter (mg-N/L) nitrate from deionized (DI) water amended with various metal ions at 1, 10, and 100 times their National Water Quality Laboratory annual median concentrations as listed in table 18. The leftmost column, labeled DI, indicates recovery of 2.50 mg-N/L nitrate from DI water not amended with metal ions. Column heights and associated error bars at tops of columns indicate the average and standard deviation of nitrate recovery for three replicate determinations of each test solution.

Table 18. Metal ions tested for possible inhibition of AtNaR2 enzyme and Griess-reaction interference.

[µg/L, microgram per liter; µM, micromole per liter]

Metal ion	Median concentration (µg/L)	Atomic weight	Median concentration (µM)	10x Median concentration (µM)	100x Median concentration (µM)
Ba^{2+}	4.07×10^1	137.34	2.96×10^{-1}	2.96×10^0	2.96×10^1
Ca^{2+}	3.64×10^4	40.08	9.08×10^2	9.08×10^3	9.08×10^4
Cr^{3+}	2.60×10^{-1}	52.00	5.00×10^{-3}	5.00×10^{-2}	5.00×10^{-1}
$Fe^{2+} + Fe^{3+}$	7.93×10^0	55.85	1.42×10^{-1}	1.42×10^0	1.42×10^1
$Li+$	7.11×10^0	6.94	1.02×10^0	1.02×10^1	1.02×10^2
Mg^{2+}	8.33×10^3	24.31	3.43×10^2	3.43×10^3	3.43×10^4
Mn^{2+}	1.07×10^1	54.94	1.95×10^{-1}	1.95×10^0	1.95×10^1
Sr^{2+}	2.7×10^2	87.62	3.04×10^0	3.04×10^1	3.04×10^2

Demonstration of Method Capability

With the properties of AtNaR2 as an analytical reagent for reducing nitrate to nitrite fully characterized, we began a four-part demonstration of capability for standard- and low-level DA nitrate + nitrite assays using the soluble AtNaR2:NADH reagent system. The DA, enzymatic-reduction assay results were in all cases compared with corresponding USGS-approved CFA-CdR assays. Part 1 confirms that for typical DA assays thermostatted at 37°C, AtNaR2 is a better reagent than YNaR1 for quantitatively reducing nitrate to nitrite. Part 2 is a graphical demonstration that analytical results for representative samples obtained with standard- and low-level DA-AtNaR2 and CFA-CdR assays are equivalent. Part 3 demonstrates that DA-AtNaR2 assay analytical response to nitrate and nitrite is equivalent. Part 4 provides results of paired t-tests and nonparametric Wilcoxon sign-rank tests that demonstrate equivalence of analytical results obtained by CFA-CdR and DA-AtNaR2 methods.

Comparison of AtNaR2 and YNaR1 Reagents in Standard- and Low-Level DA Nitrate + Nitrite Assays

We analyzed a set of 115 samples—72 surface water, 32 groundwater, and 11 blind field and laboratory QC—on the DA platform, first with YNaR1 and then with AtNaR2 as the enzyme reagent. *Figure 12* provides a graphical comparison of nitrate concentrations resulting from nitrate-reductase methods in relation to those from USGS-approved CFA-CdR methods. The CFA-CdR method (NWQL laboratory code 1975) concentration data were products of routine operations in the NWQL Nutrients Unit. The inset in *figure 12* pertains to two surface-water samples and five groundwater samples with nitrate + nitrite concentrations greater than the 5 mg-N/L calibration limit of CFA and DA assays, therefore requiring manual (CFA) or online automatic (DA) dilution prior to reanalysis. In *figure 12*, nitrate + nitrite concentrations determined by the reference CFA-CdR method (x-axis) and by DA-AtNaR2 methods

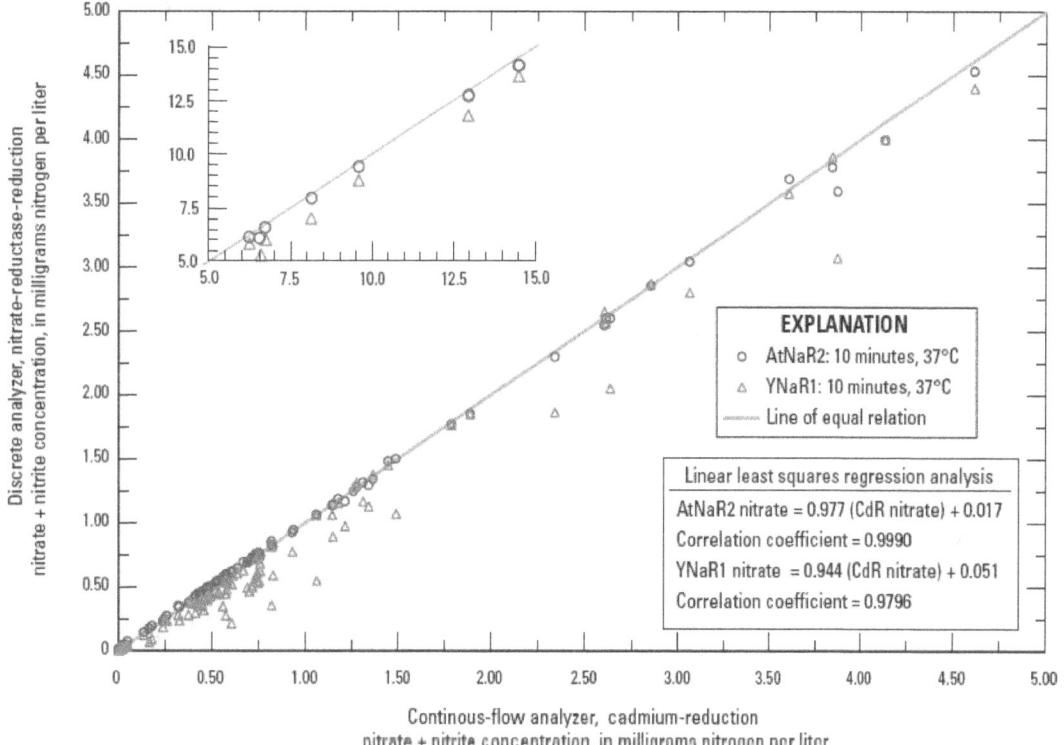

Figure 12. Comparison of nitrate + nitrite concentrations for 115 environmental water samples analyzed by automated continuous-flow, cadmium-reduction (CdR) method and automated discrete analyzer (DA) enzymatic-reduction method with NAD(P)H:YNaR1 nitrate reductase and NADH:AtNaR2 nitrate reductase. DA determinations were consecutive on November 9, 2005.

(y-axis) are plotted about the line of equal relation (slope = 1). As predicted from prior experimental results, nitrate concentrations determined by the DA method using AtNaR2 reagent (blue circles) agreed more closely with those determined by CFA-CdR reference method than did those determined by the DA method using YNaR1 reagent (red triangles). This was also the case for diluted samples (see inset graph, *fig. 12*).

Graphical Comparison of Standard- and Low-Level DA-AtNaR2 and CFA-CdR Assays

We analyzed data plotted in *figure 13* in May 2008. As identified in the figure legend, these data are noteworthy because of their wide concentration distribution and matrix diversity. Agreement between nitrate + nitrite concentrations determined by standard-level DA-AtNaR2 and USGS-approved CFA-CdR methods are excellent as indicated by regression parameters in the figure inset.

Figure 14 shows that nitrate + nitrite concentrations determined by the low-level DA-AtNaR2 method also compare well with those determined by the corresponding USGS-approved low-level CFA-CdR method. Four of the eight blind blanks for the DA-AtNaR2 method included in this figure had concentrations near the anticipated 0.008 mg-N/L interim reporting limit, but they are within the NWQL Nutrients Unit blank concentration criteria of ±1 interim or long-term MDL.

Equivalence of Low-Level DA-AtNaR2 Assay Response to Nitrate and Nitrite

As with the CFA-CdR assays they replace, DA-AtNaR2 assays measure the sum of nitrate and nitrite. Nitrate concentrations, therefore, are calculated by subtraction of independently determined nitrite concentrations. Summary statistics for nitrite concentrations determined at the NWQL in 2004 and 2010 provided in *tables 19* and *20*, respectively, demonstrate that nitrite concentrations typically are small. For example, with reference to *table 20*, 95 percent of surface-water and groundwater samples analyzed for nitrate at the NWQL in 2010 had nitrite concentrations less than 0.06 mg-N/L. Furthermore, 21 percent of the 8,418 surface-water samples and 69 percent of the 2,719 groundwater samples analyzed had nitrite concentrations less than the 0.001 mg-N/L MDL. *Figure 15* demonstrates that the small population of samples analyzed at the NWQL in 2010 containing nitrite concentrations of 0.05 mg-N/L or more, nitrate was predominant without exception and nitrite concentrations did not exceed 0.2 mg-N/L up to the 5.0 mg-N/L dilution limit of the standard-level nitrate + nitrite assay. Because of the demonstrated low nitrite concentration in surface water and groundwater, we evaluated response of the low-level DA-AtNaR2 nitrate + nitrite assay to nominally identical concentrations of nitrate and nitrite. In one experiment, we prepared DI water solutions

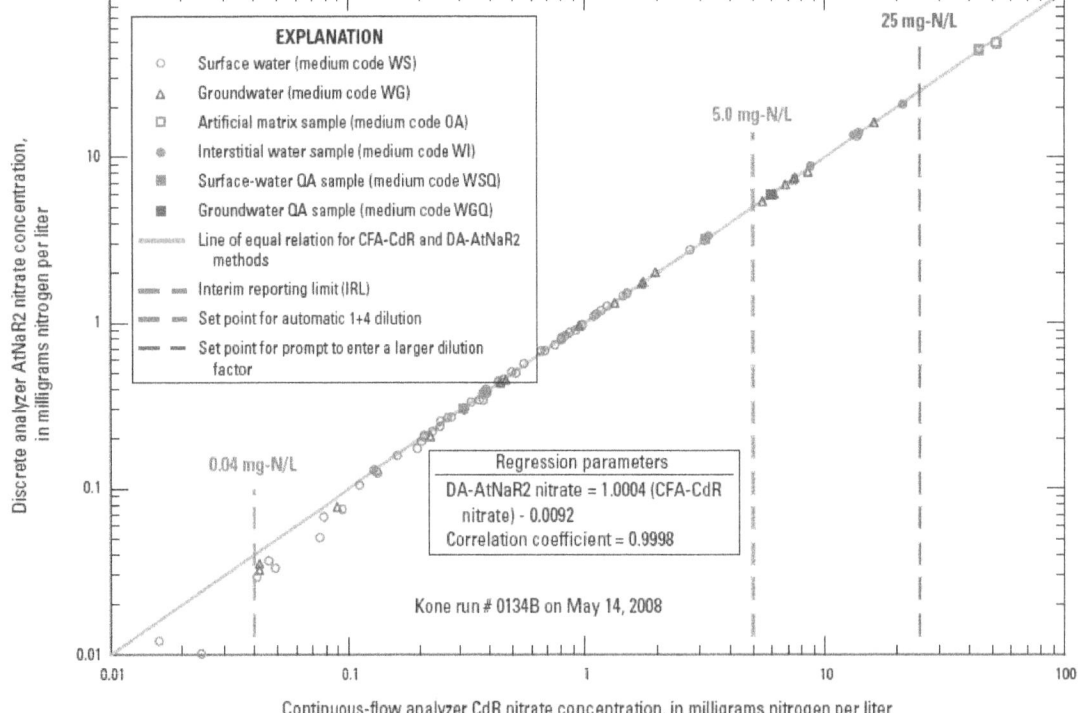

Figure 13. Comparison of nitrate + nitrite concentrations of 101 environmental water samples analyzed by standard-level automated continuous-flow analyzer, cadmium-reduction method (CFA-CdR) and standard-level automated discrete analyzer, AtNaR2-nitrate-reductase reduction method (DA-AtNaR2). (QA, quality assurance)

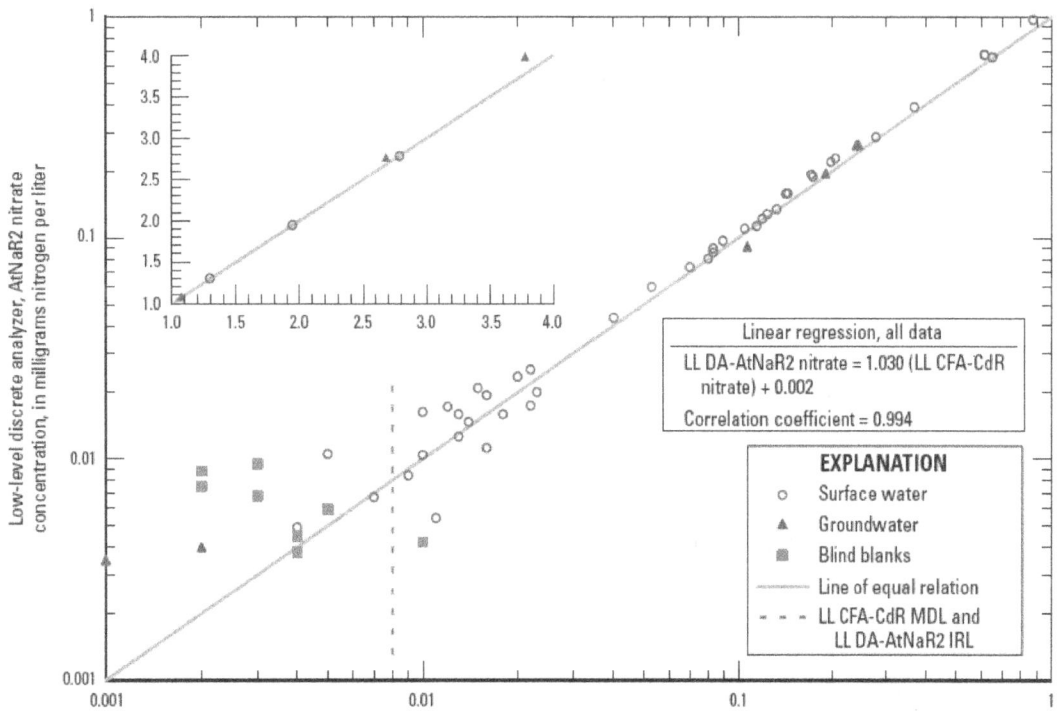

Figure 14. Comparison of nitrate + nitrite concentrations for 67 environmental water samples analyzed by low-level (LL) automated continuous-flow analyzer, cadmium-reduction method (CFA-CdR) and low-level automated discrete analyzer, AtNaR2-nitrate-reductase reduction method (DA-AtNaR2). One point (0.994, 1.043) included in linear regression analysis is not shown. (MDL, method detection limit; IRL, interim reporting limit)

Table 19. Summary statistics for nitrite concentrations determined in surface water and groundwater during calendar year 2004 at the National Water Quality Laboratory (NWQL) by automated continuous-flow analyzer methods.

[MDL, method detection limit; mg NO_2^--N/L, milligram nitrite nitrogen per liter; DL, dilution limit; WG, groundwater medium code; WS, surface water medium code; <, less than; ≤, less than or equal to]

NWQL laboratory code (test name)	1973 (Standard-level nitrite)		1977 (Low-level nitrite)	
MDL (mg NO_2^--N/L)	0.004		0.001	
DL (mg NO_2^--N/L)	1.000		0.200	
Medium code	WG	WS	WG	WS
Total samples analyzed	3,842	6,439	210	1,932
Counts: Concentration < MDL (percent of total)	3,100 (81)	2,583 (40)	116 (55)	458 (24)
Counts: MDL ≤ Concentration ≤ DL (percent of total)	739 (19)	3,845 (60)	94 (45)	1,471 (76)
Counts: Concentration > DL (percent of total)	3 (<1)	11 (<1)	0	3 (<1)
50th percentile concentration (mg NO_2^--N/L)	0.012	0.014	0.002	0.003
75th percentile concentration (mg NO_2^--N/L)	0.030	0.033	0.004	0.007
95th percentile concentration (mg NO_2^--N/L)	0.169	0.124	0.007	0.025
Maximum concentration (mg NO_2^--N/L)	1.29	6.61	0.025	0.045

Table 20. Summary statistics for nitrite concentrations determined in surface water and groundwater during calendar year 2010 at the National Water Quality Laboratory (NWQL) by the automated discrete analyzer method.

[MDL, method detection limit; mg NO_2^--N/L, milligram nitrite nitrogen per liter; DL, dilution limit; WG, groundwater medium code; WS, surface water medium code; <, less than; ≤, less than or equal to]

NWQL laboratory code (test name)	3117 (Nitrite)	
MDL (mg NO_2^--N/L)	0.001	
DL (mg NO_2^--N/L)	0.200	
Medium code	WG	WS
Total samples analyzed	2,719	8,418
Counts: Concentration < MDL (percent of total)	1,880 (69)	1,793 (21)
Counts: MDL ≤ Concentration ≤ DL (percent of total)	839 (31)	6,623 (79)
Counts: Concentration > DL (percent of total)	0	2 (<0.1)
50th percentile concentration (mg NO_2^--N/L)	0.002	0.005
75th percentile concentration (mg NO_2^--N/L)	0.008	0.014
95th percentile concentration (mg NO_2^--N/L)	0.055	0.059
Maximum concentration (mg NO_2^--N/L)	0.778	2.018

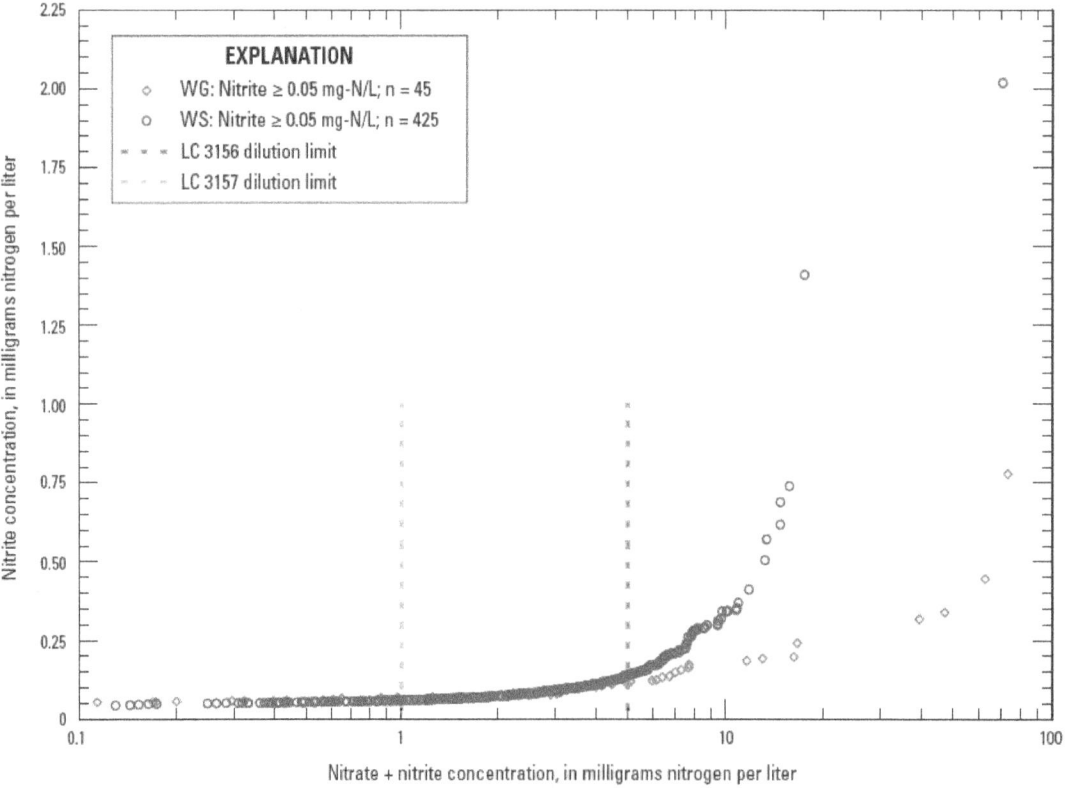

Figure 15. The relation between nitrate + nitrite and nitrite concentrations for surface-water (medium code WS) and groundwater (medium code WG) samples analyzed at the National Water Quality Laboratory in 2010 with nitrite concentrations of 0.05 mg-N/L or more. (mg-N/L, milligram nitrogen per liter; n, number of samples; LC, laboratory code)

of nitrite in the concentration range of 0.00 mg-N/L to 0.90 mg-N/L and analyzed them on the DA platform for nitrite (NWQL laboratory code 3117) and low-level nitrate + nitrite (NWQL laboratory code 3157). In another, we prepared DI water solutions containing nitrite and nitrate in six concentration ratios that nominally summed to 0.50 mg-N/L and again analyzed them on the DA for nitrite and nitrate + nitrite by NWQL laboratory codes 3117 and 3157, respectively. Calibrants for the low-level DA-AtNaR2 nitrate + nitrite assay contained only nitrate. *Table 21* provides nominal concentrations of test solutions and analytical results for both sets of experiments, which demonstrate near equivalent response of the low-level DA-AtNaR2 nitrate assay to nitrate and nitrite individually and in combination.

Statistical Comparisons of Nitrate + Nitrite Results Determined by CFA-CdR and DA-AtNaR2 Methods

Summary statistics for nitrate + nitrite concentrations determined in filtered-water samples by CFA-CdR and DA-AtNaR2 methods appear in *table 22*. Box plots in *figure 16* show distributions of standard-level nitrate + nitrite concentrations determined by CFA-CdR and DA-AtNaR2 methods in surface water (WS) and groundwater (WG) for 476 samples included in validation experiments between November 2005 and May 2008. Paired t-test analysis (Pollard, 1979) of all standard-level data (539 data pairs; see *table 23*)

indicates that the difference between means of cadmium- and AtNaR2-reduction method nitrate concentrations are not statistically different from zero at the 0.05 or 0.01 probability level. Furthermore, differences between population means for surface-water, groundwater, and "other" subsets of these data are less than the NWQL MDL for the standard-level CFA-CdR nitrate method (NWQL laboratory code 1975) and are therefore analytically insignificant. When paired t-test analyses are restricted to "in-range" data (454 data pairs with concentrations less than or equal to 5.00 mg NO_3^--N/L), differences between population means for "all data" and associated surface water and groundwater subsets are statistically different from zero at the 0.05 and 0.01 probability levels. However, differences between population means are not analytically significant. It is also noteworthy that DA-AtNaR2 assay population means were equal to or slightly greater than nitrate concentrations measured by USGS-approved CFA-CdR reference assays. This result is in sharp contrast to nitrate concentrations measured by DA-YNaR1 assays that on average were biased low in relation to CFA-CdR assays (see *figs. 8* and *12*). Nonparametric Wilcoxon signed-rank test (Pollard, 1979) results for these data (see *table 24*) are in general agreement with paired t-test results; that is, differences in nitrate concentration populations measured by CFA-CdR and DA-AtNaR2 assays are statistically significant at the 0.05 and 0.01 probability levels, but calculated differences between population medians (*table 22*) are less than the 0.02 mg-N/L MDL and are therefore not analytically significant.

Table 21. Data demonstrating near equivalent response of low-level (LL) AtNaR2 nitrate + nitrite assay (National Water Quality Laboratory laboratory code 3157) to nitrate and nitrite individually and combined. Calibrants for the LL DA-AtNaR2 nitrate + nitrite assay contained only nitrate.

[mg-N/L, milligram nitrogen per liter; NO_2^-, nitrite; NO_3^-, nitrate; SD, standard deviation; n, number of samples]

Nominal (mg-N/L)		LL NO_3^- + NO_2^- (mg-N/L)		NO_2^- (mg-N/L)	
NO_2^--N	NO_3^--N	Found	SD (n = 3)	Found	SD (n = 3)
0.00	0.00	0.01	0.001	0.001	0.0000
0.05	0.00	0.06	0.010	0.050	0.0003
0.10	0.00	0.10	0.001	0.101	0.0004
0.15	0.00	0.15	0.000	0.152	0.0002
0.30	0.00	0.29	0.001	0.293	0.0053
0.60	0.00	0.58	0.004	0.600	0.0003
0.90	0.00	0.87	0.009	0.912	0.0015
0.00	0.50	0.50	0.005	0.001	0.0000
0.05	0.45	0.49	0.001	0.050	0.0003
0.10	0.40	0.49	0.001	0.098	0.0002
0.15	0.35	0.49	0.002	0.151	0.0002
0.20	0.30	0.49	0.003	0.190	0.0027
0.25	0.25	0.49	0.003	0.242	0.0045

Table 22. Summary statistics for nitrate + nitrite concentrations determined in filtered-water samples by continuous-flow analyzer, cadmium-reduction (CFA-CdR) method and discrete analyzer, nitrate-reductase-reduction (DA-AtNaR2) method.

[mg-N/L, milligram nitrogen per liter; n, number of samples; MC, U.S. Geological Survey sample medium code; WG, groundwater sample medium code (formerly 6); WS, surface-water sample medium code (formerly 9); NWQL, National Water Quality Laboratory]

	n	Minimum	25th percentile	Median	75th percentile	Maximum
Standard-level CFA-CdR (NWQL laboratory code 1975) and DA-AtNaR2 nitrate methods						
All CdR	539	−0.003	0.367	0.922	2.815	77.3
All AtNaR2	539	−0.028	0.368	0.922	2.854	75.5
MC WG CdR	238	−0.003	0.527	1.281	4.126	77.3
MC WG AtNaR2	238	−0.028	0.525	1.289	4.353	75.5
MC WS CdR	238	0.000	0.415	0.731	1.986	14.3
MC WS AtNaR2	238	−0.025	0.440	0.732	1.987	14.9
MC (other)[1] CdR	63	−0.002	0.011	0.043	1.65	14.2
MC (other)[1] AtNaR2	63	−0.027	0.002	0.031	1.72	14.4
Standard-level CFA-CdR and DA-AtNaR2 nitrate methods (in-range results)[2]						
All CdR	454	−0.003	0.225	0.714	1.49	5.00
All AtNaR2	454	−0.028	0.222	0.714	1.52	4.96
MC WG CdR	185	−0.003	0.347	0.912	1.92	5.00
MC WG AtNaR2	185	−0.028	0.338	0.922	2.00	4.96
MC WS CdR	212	0.000	0.370	0.653	1.31	4.97
MC WS AtNaR2	212	−0.025	0.371	0.639	1.32	4.82
MC (other)[1] CdR	57	−0.002	0.009	0.018	1.02	4.09
MC (other)[1] AtNaR2	57	−0.027	0.0015	0.016	1.04	4.21
Low-level CFA-CdR (NWQL laboratory code 1979) and DA-AtNaR2 nitrate methods						
All CdR	67	−0.002	0.010	0.070	0.205	3.78
All AtNaR2	67	−0.011	0.010	0.074	0.229	3.99
MC WG CdR	9	0.001	0.054	0.241	1.88	3.78
MC WG AtNaR2	9	0.004	0.048	0.263	1.928	3.99
MC WS CdR	46	0.004	0.014	0.075	0.200	2.79
MC WS AtNaR2	46	−0.011	0.016	0.077	0.223	2.78
MC (other)[1] CdR	12	−0.002	0.002	0.004	0.109	0.616
MC (other)[1] AtNaR2	12	0.002	0.004	0.0075	0.121	0.673

[1] Typically laboratory-blind quality control samples and blanks.

[2] Samples with nitrate + nitrate concentrations greater than 5 mg-N/L that required dilution are not included.

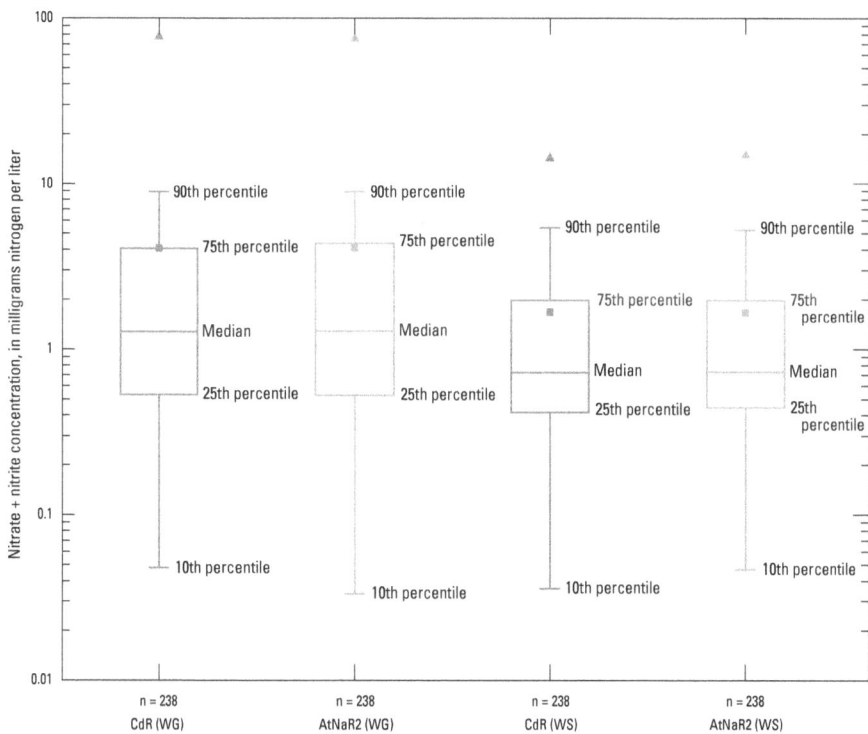

Figure 16. Concentration distributions for the population of 476 standard-level nitrate + nitrite concentrations determined in surface water (WS) and groundwater (WG) by cadmium-reduction (CdR), continuous-flow analysis and nitrate reductase (AtNaR2), discrete analysis for validation experiments spanning November 2005 through May 2008. Square and triangle symbols in each box plot indicate mean and maximum concentrations, respectively. (n, number of samples)

Table 23.   Two-population paired t-test results for nitrate + nitrite concentrations determined in filtered-water samples by continuous-flow analyzer, cadmium-reduction (CFA-CdR) method and discrete analyzer, nitrate-reductase-reduction (DA-AtNaR2) method.

[mg-N/L, milligram nitrogen per liter; n, number of samples; MC, U.S. Geological Survey sample medium code; WG, groundwater sample medium code (formerly 6); WS, surface water sample medium code (formerly 9); NWQL, National Water Quality Laboratory]

	n	Nitrate + nitrite concentration (mg-N/L)					Significant[1]		
		Mean		Difference	Standard deviation				
		CdR	AtNaR2	CdR – AtNaR2	CdR	AtNaR2	p_{calc}	$p_{0.05}$	$p_{0.01}$
Standard-level CdR (NWQL laboratory code 1975)									
All	539	2.715	2.721	−0.005	6.47	6.35	0.564	no	no
MC WG	238	4.073	4.085	−0.012	9.18	8.99	0.555	no	no
MC WS	238	1.675	1.671	0.004	2.31	2.29	0.504	no	no
MC (other)[2]	63	1.514	1.523	−0.018	2.85	2.90	0.119	no	no
Standard-level CdR (in-range results)[3]									
All	454	1.085	1.100	−0.015	1.15	1.17	0.0000	yes	yes
MC WG	185	1.31	1.34	−0.024	1.25	1.29	0.0000	yes	yes
MC WS	212	0.986	0.994	−0.008	1.004	1.003	0.0010	yes	yes
MC (other)[2]	57	0.722	0.729	−0.007	1.175	1.208	0.3455	no	no
Low-level CdR (NWQL laboratory code 1979)									
All	67	0.332	0.344	−0.012	0.710	0.731	0.0040	yes	yes
MC WG	9	0.925	0.962	−0.038	1.374	1.442	0.1635	no	no
MC WS	46	0.282	0.289	−0.007	0.552	0.555	0.0104	yes	no
MC (other)[2]	12	0.078	0.087	−0.009	0.178	0.194	0.0762	no	no

[1] p_{calc} is the probability that population means of nitrate + nitrite concentrations determined by the CdR- and YNaR1-reduction methods are the same—that is, difference between the population means is statistically equivalent to zero—on the basis of calculated paired t-tests. Difference between population means is significant at the 95 percent confidence level ($p_{0.05}$) when p_{calc} is less than 0.05 and at the 99 percent confidence level ($p_{0.01}$) when p_{calc} is less than 0.01.

[2] Typically laboratory-blind quality control samples and blanks.

[3] Samples with nitrate + nitrate concentrations greater than 5 mg-N/L that required dilution are not included.

Table 24. Wilcoxon signed-rank test[1] results for nitrate + nitrite concentrations determined in filtered-water samples by continuous-flow analyzer, cadmium-reduction (CFA-CdR) method and discrete analyzer, nitrate-reductase-reduction (DA-AtNaR2) method.

[mg-N/L, milligram nitrogen per liter; n, number of samples; MC, U.S. Geological Survey sample medium code; WG, groundwater sample medium code (formerly 6); WS, surface water sample medium code (formerly 9); NWQL, National Water Quality Laboratory; <, less than]

AtNaR2-CdR	Positive ranks	Negative ranks	Methods results populations different?[2]		
			p_{calc}	$p_{0.05}$	$p_{0.01}$
Standard-level CdR samples (NWQL laboratory code 1975)					
All MCs	348	191	< 0.0001	yes	yes
MC WG	148	90	0.0002	yes	yes
MC WS	170	68	< 0.0001	yes	yes
MC (other)[3]	30	33	0.8011	no	no
Standard-level CdR samples (in-range results)[4]					
All MCs	304	150	< 0.0001	yes	yes
MC WG	116	69	0.0007	yes	yes
MC WS	162	50	< 0.0001	yes	yes
MC (other)[3]	26	31	0.5962	no	no
Low-level CdR samples (NWQL laboratory code 1979)					
All MCs	50	17	< 0.0001	yes	yes
MC WG	7	2	0.1797	no	no
MC WS	33	13	0.0051	yes	yes
MC (other)[3]	10	2	0.0386	yes	no

[1] The paired-sample, Wilcoxon signed-rank test is a nonparametric alternative to the paired-sample t-test. It can be used to examine whether or not two paired sample populations have the same distribution. Unlike the paired-sample t-test, this function does not require either test population to be normally distributed.

[2] p_{calc} is the probability that population distributions of nitrate + nitrite concentrations determined by the CdR- and YNaR1-reduction methods are the same on the basis of the paired sample, Wilcoxon signed-rank test. The difference between population distributions is significant at the 95 percent confidence level $(p_{0.05})$ when p_{calc} is less than 0.05 and at the 99 percent confidence level $(p_{0.01})$ when p_{calc} is less than 0.01.

[3] Typically laboratory-blind quality control samples and blanks.

[4] Samples with nitrate + nitrate concentrations greater than 5 mg-N/L that required dilution are not included.

Conclusions

Numbered conclusions in the list that follows correspond to the list of objectives in the section "Purpose and Scope."

1. Paired statistical and graphical analyses of nitrate + nitrite concentrations determined in more than 500 seasonally, geographically, and composition-ally diverse surface-water and groundwater samples demonstrate the comparability of analytical results determined by standard- and low-level continuous-flow analyzer, cadmium-reduction (DFA-CdR) and discrete analyzer, nitrate-reductase-reaction (DA-AtNaR2) methods. Effects on nitrate + nitrite con-centration trend analysis across this method-change boundary should be negligible.

    a. Paired t-test statistical analyses of results from-CFA-CdR and DA-AtNaR2 methods (see *table 23*)

indicate that the difference between popula-tion means of nitrate + nitrite concentrations determined by the two methods was statistically equivalent to zero at the 0.05 probability level. Nonparametric Wilcoxon signed-rank statistical analyses (see *table 24*) indicate that the concentra-tion distributions of the same population are not equally distributed at the 0.05 probability level. With reference to summary statistics in *table 22* and box plots in *figure 16*, however, the differ-ence between population medians is less than the method detection limit (MDL) (0.02 mg-N/L) and therefore not analytically significant. Although the difference between means for the subset of surface-water samples that did not require dilu-tion (n = 212) was statistically different than zero, the calculated difference (-0.008 mg-N/L) is not analytically different from zero. The difference between means for the subset of groundwater samples that did not require dilution (n = 185) also was statistically different than zero, and the calculated difference (-0.024 mg-N/L) is analytically significant. In this case, however, the population mean difference is negative—that is, DA-AtNaR2 assay nitrate + nitrite concentrations were on average slightly greater than those for the CFA-CdR assay (see *table 23*). Trends evident in graphical analysis and calculated linear least-squares regression parameters within the body of this report support these results.

b. Paired t-test statistical analyses of results from low-level CFA-CdR and DA-AtNaR2 methods (see *table 23*) indicate the difference between nitrate concentration population means determined by the two methods were statistically different from zero at the 0.05 probability level. This statistically significant difference between populations means (-0.012 mg-N/L), however, is negative—that is, low-level DA-AtNaR2 assay nitrate + nitrite concentrations were on average slightly greater than those for the low-level CFA-CdR assay. This was also the case for nonparametric Wilcoxon signed-rank statistical analyses (see *table 24*), which indicate that the concentration distributions of the two populations are not equally distributed at the 0.05 probability level. With reference to *table 22*, differences between population medians are not analytically significant. Differences in calculated population means (*table 23*) and medians (*table 22*) for groundwater samples (-0.038 mg-N/L and -0.022 mg-N/L, respectively) were statistically and analytically significant. This might be due in part to the small number of samples in this population (n = 9), but it might also reflect better tolerance to reduced metals and sulfides by the enzymatic-reduction assays than the cadmium-reduction assays. As was the case for standard-level population mean differences, low-level method differences were negative—that is, AtNaR2-reduction method nitrate concentrations on average were slightly greater than cadmium-reduction method nitrate concentrations (see *table 23*). Trends evident in graphical analysis and calculated linear least-squares regression parameters within the body of this report support these results.

2. Complete operational details (preparation of reagents, calibrants, and QC solutions) and performance benchmarks for these new methods (MDLs, blank levels, between-day precision, and spike recovery) are provided for analysts at the NWQL and elsewhere who need to implement these methods and operate them routinely.

3. Experimental results provided in this report demonstrate negligible interference in either enzymatic or colorimetric assay reaction steps by common surface-water and groundwater matrix constituents, such as major and minor ions and humic substances, over a reaction temperature range of 5°C to 37°C.

   a. Anions and cations at concentrations up to 100 times their median concentrations in typical freshwater matrices have negligible effects on the activity of AtNaR2 nitrate reductase. Group II cations suppress formation of Griess reaction chromophore. Calcium ions at NWQL-median concentrations exert the greatest suppression, although barium ions are more potent indicator reaction suppressors on the basis of molar concentration. Due to thermal instability of nitrous acid and diazonium intermediates (Noller, 1966), both the yield of Griess indicator reaction chromophore and its formation rate are inversely proportional to reaction temperature in the range of 10°C to 50°C.

   b. High-phenolic-content humic acids (HAs) do not inhibit AtNaR2 at reaction temperatures ranging between 5°C and 37°C. This unique property makes AtNaR2 the reagent of choice for DA nitrate determination methods described in this report and other natural water nitrate assays that are most easily performed at or above typical ambient laboratory temperatures (72–78°F, 22–26°C).

Acknowledgments

We are grateful to Ms. Christine McEvoy for technical assistance in summer 2003. At that time, she was a chemistry major at the College of Wooster in Wooster, Ohio, and her participation in our research was supported through the USGS Summer Student Intern Program. Authors Kryskalla and Patton thank the Nitrate Elimination Company in Lake Linden, Mich., for partially funding their 2003 and 2004 research at the NWQL through a U.S. Department of Agriculture Phase-II small business innovation research (SBIR) grant (USDA SBIR # 2002-33610-12300, Environmentally benign automated nitrate analysis) and for continuing collaborative support and providing NADH:nitrate reductase from *Arabidopsis thaliana* in 2005 and 2006. We also acknowledge Harold Ardourel, Eric Schwab, Mary Olson, Chris Klimper, Colleen Gupta (retired), and other members of the NWQL Analytical Services Nutrients Unit for their interest and technical support during development of automated DA nitrate methods described in this report.

The report also benefited from reviews by Nancy Simon at the USGS National Research Program in Reston, Va., Carl Zimmermann and Carolyn Keefe at the University of Maryland Chesapeake Biological Laboratory in Solomons, Md., and others at the USGS Office of Water Quality and the NWQL.

References Cited

American Society for Testing and Materials, 1999, Reporting test results, section 7.4 *of* ASTM E29-93a, Standard practice for using significant digits in test data to determine conformance to specifications, *in* General test methods, forensic sciences, terminology, conformity assessment, statistical methods. Annual book of ASTM standards, v. 14.02: West Consohocken, Pa., ASTM International, 4 p.

American Society for Testing and Materials, 2001, ASTM D1193-99, Standard specification for reagent water, *in* Water (I). Annual book of ASTM standards, v. 11.01: West Consohocken, Pa., ASTM International, p. 107–109.

Bendschneider, Kenneth, and Robinson, R.J.,1952, A new spectrophotometric method for the determination of nitrite in sea water: Journal of Marine Research, v. 11, p. 87–96.

Bratton, A.C., and Marshall, E.K., 1939, A new coupling component for sulfanilamide determination: Journal of Biological Chemistry, v. 128, p. 537–550.

Campbell, W.H., Song, Pengfei, and Barbier, G.G., 2006, Nitrate reductase for nitrate analysis in water: Environmental Chemistry Letters, v. 4, p. 69–73.

Colman, B.P., and Schimel, J.P., 2010a, Understanding and eliminating iron interference in colorimetric nitrate and nitrite analysis: Environmental Monitoring and Assessment, v. 165, p. 633–641.

Colman, B.P., and Schimel, J.P., 2010b, Erratum to—Understanding and eliminating iron interference in colorimetric nitrate and nitrite analysis: Environmental Monitoring and Assessment, v. 165, p. 693.

Davison, W., and Woof, C., 1978, Comparison of different forms of cadmium as reducing agents for the batch determination of nitrate: Analyst, v. 103, p. 403–406.

Draper, N.R., and Smith, Harry, 1966, Applied regression analysis: New York, Wiley, 407 p.

Elliot, C.L., Snyder, G.H., and Cisar, J.L., 1989, A modified AutoAnalyzer II method for the determination of NO_3-N in water using a hollow-Cd reduction coil: Communications in Soil Science and Plant Analysis, v. 20, p. 1873–1879.

Fox, J.B., 1979, Kinetics and mechanisms of the Griess reaction: Analytical Chemistry, v. 51, p. 1493–1502.

Fox, J.B., 1985, The determination of nitrite—A critical review: CRC Critical Reviews in Analytical Chemistry, v. 15, p. 283–313.

Gal, Carmen, Frenzel, Wolfgang, and Möller, Jürgen, 2004, Re-examination of the cadmium reduction method and optimisation of conditions for the determination of nitrate by flow injection analysis: Microchimica Acta, v. 146, p. 155–164.

Green, R.B., and Foreman, W.T., 2005, Guidelines for method validation and publication at the National Water Quality Laboratory [unpublished document]: National Water Quality Laboratory Standard Operating Procedure MX0015.3.

Guevara, Ibeth; Iwanejko, Joanna; Dembinska-Kiec, Aldona; Pankiewicz, Joanna; Wanat, Alicja; Anna, Polus; Golabek, Iwona; Bartus, Stanislaw; Malczewska-Malec, Malgorzata; and Szczudlik, Andrzej, 1998, Determination of nitrite/nitrate in human biological material by the simple Griess reaction: Clinica Chimica Acta, v. 274, p. 177–188.

Gupta, C.A., Patton, C.J., Kryskalla, J.R., Schwab, E.A., and Olson, M.C., 2010, Colorimetric determination of ammonium, nitrate plus nitrite, nitrite, and orthophosphate in water by automated discrete analysis [unpublished document]: National Water Quality Laboratory Standard Operating Procedure INCF0452.2.

Hansen, W.R., ed., 1991, Suggestions to authors of the reports of the United States Geological Survey (7th ed.): U.S. Government Printing Office, p. 119–121.

Ingle, J.D., and Crouch, S.R., 1988, Spectrochemical analysis: New Jersey, Prentice Hall, p. 129–130.

MacKown, C.T., and Weik, J.C., 2004, Comparison of laboratory and quick-test methods for forage nitrate: Crop Science, v. 44, p. 218–226.

Miranda, Katrina, Espey, M.M., and Wink, D.A., 2001, A rapid, simple spectrophotometric method for simultaneous detection of nitrate and nitrite: Nitric Oxide—Biology and Chemistry, v. 5, no. 1, p. 62–71.

Moody, J.A., and Shaw, F.L., 2006, Reevaluation of the Griess reaction—How much of a problem is interference by nicotinamide nucleotides?: Analytical Biochemistry, v. 356, p. 154–156.

Moorcroft, M.J., Davis, James, and Compton, R.G., 2001, Detection and determination of nitrate and nitrite—A review: Talanta, v. 54, p. 785–803.

Mori, Hisakazu, 2000, Direct determination of nitrate using nitrate reductase in a flow system: Journal of Health Science, v. 46, no. 5, p. 385–388.

Mori, Hisakazu, 2001, Determination of nitrate in biological fluids using nitrate reductase in a flow system: Journal of Health Science, v. 47, no. 1, p. 65–67.

Noller, C.R., 1966, Chemistry of organic compounds, (3d ed.): Philadelphia, W.B Saunders Company, 542 p.

Norwitz, George, and Keliher, P.N., 1985, Study of interferences in the spectrophotometric determination of nitrite using composite diazotization-coupling reagents: Analyst, v. 110, p. 689–694.

Norwitz, George, and Keliher, P.N., 1986, Study of organic interferences in the spectrophotometric determination of nitrite using composite diazotization-coupling reagents: Analyst, v. 111, p. 1033–1037.

Novak, C.E., 1985, Preparation of water-resources data reports: U.S. Geological Survey Open-File Report 85–480, 331 p.

Nydahl, Folke, 1976, On the optimum conditions for the reduction of nitrate to nitrite by cadmium: Talanta, v. 23, 349–357.

Pai, Su-Cheng, Yang, Chung-Cheng, and Riley, J.P., 1990, Formation kinetics of the pink azo dye in the determination of nitrite in natural waters: Analytica Chimica Acta, v. 232, p. 345–349.

Patton, C.J., 1983 Design, characterization, and applications of a miniature continuous flow analysis system—Ph.D. dissertation, Michigan State University [abs.]: Dissertation Abstracts International, v. 44, p. 778–B.

Patton, C.J., Fischer, A.E., Campbell, W.H., and Campbell, E.R., 2002, Corn leaf nitrate reductase—A nontoxic alternative to cadmium for photometric nitrate determinations in water samples by air-segmented continuous-flow analysis: Environmental Science and Technology, v. 36, p. 729–735.

Patton, C.J., and Gilroy, E.J., 1998, U.S. Geological Survey Nutrient Preservation Experiment—Experimental design, statistical analysis, and interpretation of analytical results: U.S. Geological Survey Water–Resources Investigations Report 98–4118, 73 p.

Patton, C.J., and Rogerson, P.F., 2007, Reduction device for nitrate determination: U.S. Patent 7,157,059 B1 [filed September 18, 2002; issued January 2, 2007].

Patton, C.J., and Truitt, E.P., 1995, U.S. Geological Survey Nutrient Preservation Experiment—Nutrient concentration data for surface-, ground-, and municipal-supply water samples and quality assurance samples: U.S. Geological Survey Open-File Report 95–141, 140 p.

Pinto, P.C.A.G., Lima, J.J.F.C., and Saraiva, M.L.M.F.S., 2005, An enzymatic analysis methodology for the determination of nitrates and nitrites in water: International Journal of Environmental Chemistry, v. 85, no. 1, p. 29–40.

Pollard, J.H., 1979, A handbook of numerical and statistical techniques: New York, Cambridge University Press.

Senn, D.R., and Carr, P.W., 1976, Determination of nitrate at the part per billion level in environmental samples with a continuous-flow immobilized enzyme reactor: Analytical Chemistry, v. 46, no. 7, p. 954–958.

Schwab, E.A., Patton, C.J., and Kryskalla, J.R., 2009, Kone Aquakem 600™ DA, version 1.0 [unpublished document]: National Water Quality Laboratory Technical Operations Manual.

Skipper, Lawrie, Campbell, W.H., Mertens, J.A., and Lowe D.J., 2001, Pre-steady-state kinetic analysis of recombinant Arabidopsis NADH nitrate reductase: Journal of Biological Chemistry, v. 276, no. 29, p. 26998–27002.

Stainton, M.P., 1974, Simple, efficient reduction column for use in the automated determination of nitrate in water: Analytical Chemistry, v. 46, p. 1616.

U.S. Environmental Protection Agency, 1997, Definition and procedures for the determination of the method detection limit, app. B *to* Guidelines establishing test procedures for the analysis of pollutants, part 136 *of* Protection of the environment—U.S. Code of Federal Regulations, Title 40: National Archives and Records Administration–Office of the Federal Register, Government Printing Office, p. 265–267.

U.S. Environmental Protection Agency, 1995, National primary drinking water regulations—Nitrates and nitrites: U.S. Environmental Protection Agency EPA 811–F–95–002f–T, 2 p.

U.S. Geological Survey, 2002, Policy for storing and reporting significant figures for chemical data: Office of Water Quality Technical Memorandum 2002.11, accessed September 6, 2005, at *http://water.usgs.gov/admin/memo/QW/qw02.11. html*.

Willis, R.B., 1980, Reduction column for automated determination of nitrate and nitrite in water: Analytical Chemistry, v. 52, p. 1376–1377.

Willis, R.B., and Gentry, C.E., 1987, Automated method for determining nitrate and nitrite in water and soil extracts: Communications in Soil Science and Plant Analysis, v. 18, p. 625–636.

Wood, E.D., Armstrong, F.A.J., and Richards, F.A., 1967, Determination of nitrate in sea water by cadmium-copper reduction to nitrite: Journal of the Marine Biological Association of the United Kingdom, v. 47, p. 23–31.

Zhang, J.Z., Fischer, C.J., and Ortner, P.B., 2000, Comparison of open tubular cadmium reactor and packed cadmium column in automated gas-segmented continuous flow nitrate analysis: International Journal of Environmental Analytical Chemistry, v. 76, p. 99–113.

www.ingramcontent.com/pod-product-compliance
Lightning Source LLC
Chambersburg PA
CBHW081359170526
45166CB00010B/3143